GALILEO

Galileo Galilei

Written by
Rebecca B. Marcus

Pictures by
Richard Mayhew

Originally published in 1961

Cover design by Robin Fight
Cover illustration by Tanya Glebova and Kessler Garrity
© 2020 Jenny Phillips
goodandbeautiful.com

For Abe

CONTENTS

I. The Great Awakening 1
II. A Cathedral Lamp Swings into Fame . . . 8
III. A Wanderer Returns. 16
IV. Experiments from a Tower 23
V. Padua 30
VI. A Master Achieves Recognition. 36
VII. Golden Years 42
VIII. A Simple Spyglass Becomes a Telescope . 50
IX. The Heavens Reveal Their Splendors . . 57
X. A Native Son Returns. 66
XI. Galileo Revisits Rome 74
XII. Out of Retirement 82
XIII. The "Dialogue" Brings Great Trouble . . 90
XIV. Truth Cannot Die. 100
XV. Important Dates in Galileo's Life 109

I

The Great Awakening

AT THE BEGINNING OF THE fifteenth century, the Western world, like a sleeping giant, had just awakened. It yawned, stretched itself, and began to look around at the civilization which met its eyes.

For many hundreds of years, Europe had been living in that period of ignorance and superstition known as the Dark Ages. Few people could read and write. Even the wealthiest lords and ladies seldom took the trouble to learn but depended upon their scribes to write letters for them. Those of the poorer classes, who might have wanted to know what was in books, rarely had the chance to find out. The only schools were in the monasteries where monks lived and worked, but only a very small number could attend these.

Books were for the select few. They were handwritten with pointed goose quills and ink on expensive parchment, each one taking months to complete. Imagine what one of them must have cost! And no wonder not very many were written!

Superstition lurked everywhere in people's minds. Witches could put an evil sign on a person; a black cat was a sign of bad fortune. Someone would die if another person far away stuck pins into an image of the victim. Many people wore charms against disease. If they fell ill in spite of the charm, magic words and spells were often tried as a cure. Certain crops had to be planted in the dark of the moon or in the full moon. These were only some of the many superstitious beliefs strongly held throughout Europe.

People were afraid of strangers, afraid of new ideas, afraid of anything different or not understood. If a person thought for himself or questioned some of these old beliefs, he did not dare voice his doubts because he might be punished or even put to death.

Scholars were so busy thinking and arguing about the next world after death, they had no time left for thinking much about the one they lived in. They would even spend days arguing about how many angels could stand on the point of a pin!

Now, at the start of the 1400s, so many new ideas were blossoming forth as people began to examine the world around them that it seemed as if knowledge were being born again. Indeed, the period from the 1400s to the 1700s has been called the *Renaissance*, a French word meaning rebirth.

Europe indeed was ripe for change, and the Crusades helped prepare the ground. In 1095, the pope called for the first of these holy wars against Muslims. The Crusaders were to recapture the city of Jerusalem from them. Many thousands, fired by religious zeal, "carried the Cross"—became Crusaders—and began marching to the East. Nobles, peasants, craftsmen, merchants, and even women and children joined the march.

For two hundred years, Crusaders moved across Europe

into Turkey and the Holy Land. They failed to capture Jerusalem, but they won another war—the war against darkness and ignorance. What they saw opened their eyes to different ways of life, and they were eager to introduce some of these new ways to replace the old ones.

The great Italian cities were the first to awaken. This was not an accident, for they were the main centers of trade. Florence, Genoa, and Venice lay in the path of the trade routes from the East. Ships and caravans from China and India brought their eagerly sought goods into these cities, where they would be sent on to the rest of Europe.

Small wonder, then, that as these strange cargoes were unloaded, people began to ask questions about the lands that produced them. Merchants returning from these lands told mesmerizing tales of beautiful, clean cities, richly dressed men and women, great works of art, and new foods. The Crusaders had started an interest in different things; other travelers fed that interest.

In the minds of many people, distrust of new ideas was giving way to a curiosity about what was still unknown. They were ready for new inventions.

The first great invention of the Renaissance was the printing press, about the year 1450, in Germany. Now if a scholar or scientist wrote a book, it no longer had to be written out word by word, one book at a time, taking months to complete. Many copies could be printed in one day. Books became cheaper and less difficult to get, and new ideas spread more easily.

Wealthy princes in the Italian cities vied with each other to see who could assemble the finest works of art and who could surround himself with the greatest poets, artists, and scientists. It became fashionable to make a handsome allowance to these creative people so that they would dedicate their works to their patron. A patron was generally a wealthy man who supported an artist so that the artist could devote full time to his work. Among the wealthiest, most generous patrons were the princes of Florence.

There were, of course, many who feared the new ways. What had been good enough for their fathers, they argued, was good enough for them. If ideas and theories had lasted for so long, they must be correct. In the universities and monasteries, scholars studied, almost without change, the scientific notions first taught by the Greek philosopher Aristotle, over seventeen hundred years earlier. His theories were considered the accepted truth and were not to be questioned.

* * * * * * * * * * * * * * * *

The first mighty blow against Aristotle's science was destined to be dealt by Galileo Galilei, born in Pisa, Italy, not far from Florence, on February 15, 1564.

Vincenzo and Giulia Galilei named their firstborn Galileo, after a famous ancestor who was still remembered in Florence as a great doctor. Perhaps in doing so, they hoped their son would follow his example and become a great physician who would bring fame and fortune once more to his family.

For the Galilei family was of the nobility—at one time wealthy but now become poor. For all his accomplishments as a lute player and composer, Vincenzo could not make a

Florence

living from music. And though he was a skilled mathematician, nobody at that time could see enough use for mathematics to pay him enough to support him and his family. Even the great University of Pisa did not have a single professor of mathematics among its teachers!

In despair, shortly after Galileo's birth, Vincenzo moved his family to nearby Florence. There he set up shop as a wool merchant.

The city of Florence was alive with fine artists, fine musicians, and talented writers. Learning of a high quality was in the very air. But Vincenzo could spare little time for any of it. He was tied to his wool shop. He was determined, however, that his son should profit from this environment.

Perhaps if Galileo found a life's work that would make him wealthy, he might even set his father free from his hated wool shop.

With this in mind, Vincenzo began early in the boy's life to talk of sending him to the University of Pisa to study medicine. But with his own bitter example constantly before him, he kept from his son any opportunity to study mathematics.

But what of Galileo himself? He was a stocky red-haired youth, alive with curiosity. It was not enough for him to be told a fact; he wanted to investigate it and to try to prove it for himself. Galileo wanted to know the "why" of things. His mind and his nimble fingers were always busy. When he was not playing his lute or painting pictures, he was inventing clever toy machines for the younger children in his family.

In all of these activities, he showed great ability. With practice, he might become a great musician. With training, he might become an artist. In time, he might even become an inventor. His father was bewildered by his son's many talents. Into what kind of life's work should he steer Galileo?

Vincenzo Galilei thought it over carefully. One of the most honored and well-paying professions was medicine. He would keep his first desire to have his son become a doctor.

To prepare for the University of Pisa, as well as to keep his mind so busy that he would have no time for "foolish things" like painting and toy making, Galileo was sent to the monastery school at Vallombrosa. So taken up was he with his studies that he pushed aside all other thoughts. Galileo's greatest satisfaction now was not in using his hands, but his mind.

Eagerly he studied philosophy and religion at

Vallombrosa. The restlessness he had often felt left him. Galileo felt at peace with himself and the world.

What better life could he find, Galileo thought, than that of a monk? Deep religious feelings stirred within him. Galileo determined to devote his life to the church.

Vincenzo was very disturbed at the thought of his son becoming a monk. Perhaps he understood Galileo well enough to suspect that his searching mind could never be happy under the discipline of life in a monastery. Perhaps he also feared that the family fortunes could not be restored if his oldest and most capable son devoted himself to a life of poverty.

Whatever his real reason might have been, Vincenzo found an excuse to bring Galileo home from Vallombrosa. An eye disease had made it necessary for the lad to keep away from books for a while, so back he came to Florence to rest and recover his health.

Vincenzo reasoned with the boy. Once again he reminded him of his great ancestor, the physician, and of the fame and fortune that awaited a good doctor. Galileo listened and agreed to try to study medicine, though he did not have much liking for it.

Before he was eighteen years old, the University of Pisa accepted Galileo as a medical student. Little did he dream that, in reality, his fame would rest elsewhere than in the medical profession.

II

A Cathedral Lamp Swings into Fame

"I CANNOT EVEN PRETEND TO MYSELF any longer that I am interested in the study of medicine," thought Galileo four years later.

They had been four difficult years. The youth had tried desperately to carry out his father's wishes and become a doctor, but he finally admitted defeat. Galileo shuddered at the very thought of entering a classroom in anatomy and listening to the professor drone out the names of parts of the body. Such things had little meaning for him.

Besides, he had become very unpopular with his fellow students as well as the professors. He did not seem to fit in with the young men who were planning to devote their lives to curing ill people.

"How can you cure the sick," he would say to the other students, "if you have never had the chance to carefully examine a sick person? We sit in the classroom and listen to the professor, but we ourselves never touch a patient. We are not even permitted to dissect a body. We can only watch the professor or his assistant do it."

Such talk was too daring for the other students to listen

to. Never had they dreamed of finding fault with the methods by which medicine had been taught for hundreds of years.

Galileo was challenging not only the teaching methods, but he was also questioning the very facts themselves.

"Why is this so?" "What would happen if we tried treating a patient this way instead of that way?"

The professors had no answer for him except, "This is the way it has always been done. We do not ask why."

"What nonsense some of the 'cures' were!" Galileo thought. He could not control his quick tongue and sarcastic wit.

Nor did his neglected work and frequent absence from classes endear him to his teachers. How could they be anything but hostile toward a student who studied little, seldom attended class, and, when he did, even tried to undermine the very foundations of medicine?

It may seem strange to us today, but in the 1580s, a medical student was required to spend considerable time learning the philosophy of Aristotle. This was actually more to Galileo's liking. He searched deeply into the ideas advanced by this ancient Greek philosopher and carefully examined statements made by him as being the absolute truth.

Galileo was puzzled. How could Aristotle make a positive statement about a scientific fact without ever having proved it? And how could scholars have blindly repeated these statements for seventeen hundred years? Galileo was not sure Aristotle was always right, and he hoped someday to find a way to disprove him.

But one ray of sunshine brightened Galileo's troubled days. He had discovered mathematics!

The Grand Duke of Tuscany had come from Florence

with his court to spend some time at his palace at Pisa. Ostilio Ricci, Court Mathematician to the Grand Duke, was giving a lesson to the pageboys of the palace—and Galileo accidentally overheard. He was excited. So this was the mathematics his father had not permitted him to study! But it was fascinating! It was reasonable! One step led logically to another—and all could be proved!

The young medical student approached Ricci and asked him questions. So intelligent were the questions that the court mathematician realized Galileo was not just curious, but deeply interested. Here was a young man who quickly grasped what was being explained to him; he was able to jump ahead to the next logical step almost before the master himself.

Ricci recognized the possibility of genius lying untapped in the mind of this young man. He undertook to teach him mathematics. Galileo eagerly read every book Ricci lent him until there was little left for him to learn from them.

A great restlessness seized Galileo once more. He had thought his years at the Vallombrosa monastery had calmed him, but now his inner peace had gone.

One day while he was aimlessly walking about in the streets of Pisa, he happened to stop in front of the great cathedral. Galileo looked at the beautiful arches as though he were seeing them for the first time. He had passed by often but never really taken much notice of them. Almost without thinking, he entered the cathedral, perhaps to look more closely at its magnificent interior, perhaps to pray. We shall never know, but we do know that as he stepped into the calm, serene quiet of the church, he was unknowingly coming to a turning point in his life.

Galileo sat down on a bench and looked around at the beautiful altar, the colorful mosaics, and the marble pillars

that had been brought from Greek and Roman ruins to build the cathedral hundreds of years ago. Suddenly something moving caught his eye. Some workmen who were making repairs on the building had set the great cathedral lamp swinging.

Fascinated, Galileo rose and watched it. Strange! It started swinging in a wide arc, but as the arc of its swing became smaller, its swinging became slower. He put the fingers of his right hand to his pulse on his left wrist, as he had been taught in medical school. To the regular beating of his pulse, he began timing the swinging of the lamp. Stranger still! No matter what the size of the arc, the *time* it took for the lamp to make one complete swing was the same! Even though after a while the swinging of the lamp slowed considerably, it made no difference in the total time it took to cover the distance of the arc. It *had* to go slower to swing through a smaller arc in the same time that it took to cover a greater one.

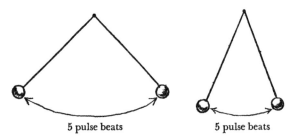

No matter what the size of the arc, the time it took for the lamp to make one complete swing was the same.

"Something was wrong," thought Galileo. Either his senses were deceiving him or Aristotle's statement that the smaller arc should take less time was incorrect. He would go back to his room and try out many different weights on different lengths of string—*pendulums*, such suspended weights were called. Only in this way could he really know the truth.

Though Galileo did not realize it, he was taking the first important step in founding the experimental science of today.

Excited, the young man returned to his lodgings. Gone was all thought of ever attending another class in medicine. Instead, he plunged feverishly into experiment after experiment with his pendulums.

He had to borrow, sometimes even beg for, the proper materials to make his experiments. String or wire or old chains were easy to obtain. Objects of the same size but of different weights were another matter. He haunted the ironmongers' shops, the carpenters' shops, and the junk shops. Finally, he laid out on his table all the materials he had obtained.

There were a number of chains, wires, and strings, two of each length. He had managed to collect pairs of iron and wooden balls, each ball of each pair being the same size but of different weights. He added an hourglass, too. Lastly, and most importantly, were his quill pen and ink and paper on which to record his observations.

Galileo set to work. So busy was he with his experiments that he rarely left his room. When the rafters to which he had attached his strings and wires proved to be too low for some of the longer pendulums, he climbed a tree in the yard outside his window and fastened a plank to a high limb. There he could hang his pendulums from greater heights for further experiments.

Finally, he was ready to make a statement. He, Galileo Galilei, had discovered by experimentation the laws governing swinging pendulums. He could calculate mathematically exactly how long it would take a pendulum to swing from one end of its arc to the other. Further, Galileo stated that, beyond a doubt, the period of time depended only upon the length of the string. No matter what the weight on the end, pendulums whose strings were of equal length swung in equal periods of time. There was no magic or mystery to it. Anyone who was interested could try the same experiment, and the results would always be the same.

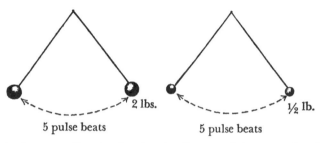

Weight does not affect the time it takes for a pendulum to swing from one end of its arc to the other.

But where could Galileo publish the results of his experiments? He had no money to print a book in which to make his claims. Nor would any printer even consider publishing anything this impulsive young man, not yet twenty years old, had to say. Besides, any such publication that went against the teachings of Aristotle would find no believers anywhere.

Instead, Galileo turned to his talents as a toymaker. He invented a "pulsifier," an instrument whereby, with the use of a small swinging pendulum, a doctor could time the pulse beat of a patient. Now, when he spoke to the professors at the University of Pisa, they listened to him respectfully. They were not much interested in Galileo's laws of the pendulum but, rather, in the practical use to which he had put them.

Galileo might have remained at the University of Pisa, very slowly gaining some recognition as a mathematician, but his father Vincenzo took a firm stand. Reports had come back to him that his son was neglecting his medical studies. It was likely that the young man would soon be asked by the authorities to leave the School of Medicine. Galileo's father demanded an explanation.

Ashamed, his son told him the truth. Though he felt very guilty in disappointing his father, never could he become a doctor. He was interested only in mathematics.

Vincenzo controlled his anger and disappointment, but he declared emphatically that he could no longer afford to keep his son at the university. His draper's shop in Florence was yielding a poor living, and there were his other children to provide for. Galileo's brother Michelangelo showed great promise as a musician, and money was needed for his further musical training. For his oldest son to remain in Pisa, in what Vincenzo considered idleness, with no preparation to earn a living, was out of the question. Galileo must come home to help out in the shop. There, perhaps, he might be able to use his mathematical ability to help straighten out the family's finances.

Sadly, as though part of his life were being torn from him, Galileo left Pisa for Florence in 1585 to start anew as a merchant's assistant.

III

A Wanderer Returns

LIFE IN THE GALILEI HOUSEHOLD in Florence was stormy. The children quarreled, their mother was quick to anger, and their father showed his discontent as a shopkeeper. And now the eldest son, Galileo, in whom so much hope had been placed, was not even earning his own living in the draper's shop.

Vincenzo sighed. So many burdens had been placed upon his shoulders! He could never please his quick-tempered, scolding, dissatisfied wife. His two growing daughters would need dowries when they married. One son wanted and deserved an expensive musical education, and another son, no longer a child, spent his days drawing strange figures on paper. This last one contributed nothing to the family. He ate food earned by others; even his shabby clothes were paid for by his father.

Mathematics! Vincenzo hated the very word. It had done him no good, and Galileo was faring even worse at it. While the father had been able to become practical and make a living of some sort, the son was still living in a dream world of mathematical figures and shapes.

Galileo himself felt he was a failure. Only the presence in Florence of Ostilio Ricci, the court mathematician, kept a spark of hope alive in his heart. He knew Ricci loved him as a son and respected him as a clear-thinking mathematician. His work with the pendulum and his solution to a number of problems—unsolved until now—so impressed Ricci that he encouraged the young man to continue his mathematical studies by himself.

"You have genius, Galileo. Be patient. In time you will be recognized properly. I speak of you often to my learned friends. 'Here is a youth,' I tell them, 'who should be watched carefully.' When I write to mathematicians all over Europe, I mention your name. Your work is better known than you think."

Ricci even spoke to Vincenzo. "Do not be too harsh with Galileo. He needs time to think. His is a greater mind than you realize. He will yet bring you honor and fame. Florence is proud of him."

"What good are these promises?" thought Vincenzo. "We need help now. This son of mine! Over twenty years old and playing with scales, pans of water, and chunks of metal."

Galileo was indeed playing with pans of water, scales, and metals, but it was not idle play. Just as he had once experimented with pendulums, he was now experimenting with the weight of objects when they were submerged in water.

Galileo had observed, as others had before him, that no two objects could be in the same place at the same time. For example, when an object was placed underwater, it pushed away, or *displaced*, an amount of water equal to its own size. If an object happened to be a cubic foot in size, it pushed away a cubic foot of water.

Moreover, the object, while underwater, weighed less than it did when in air. Strangely, the loss in weight *always* was equal to the *weight* of the displaced water.

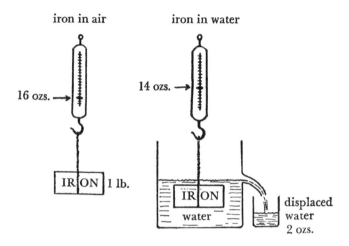

Each material, Galileo knew, had its own special size for each pound of that material. For instance, one pound (16 ounces) of pure iron always took up the same amount of space. When uncertain what a metal was made of, he could take a piece of it and weigh it in air and again while it was underwater. Then he could figure the loss of weight, which of course would tell him how much water had been displaced. Once he knew that, he could look at a chart and find out what the substance was made of.

The Greek scientist Archimedes had discovered these and other facts while trying to find out for his king whether a newly made crown was made of pure gold or not. The king, suspecting that the goldsmith had mixed the crown with cheaper metal in order to keep some gold for himself, gave Archimedes the job of testing to see. Further, said the king, he must do this without damaging the crown.

Archimedes puzzled over this for many days. At last he saw how the problem could be solved. First, he weighed the crown exactly. Next, he took a quantity of gold and a quantity of silver, each of which was the same weight as the crown. But since gold is heavier than silver, the quantity of silver was bigger in size.

Then Archimedes filled a large bowl of water right to the brim. He put the gold into the bowl and measured how much water overflowed. Then he did the same with the silver. The silver, bigger in size, made more water flow over the brim. In other words, it displaced more water because of its larger size.

If, reasoned Archimedes, the crown was really made of pure gold, it would make the same amount of water flow over the brim as the lump of gold. He put the crown in to see. But Archimedes found that the crown made *more* water overflow! The problem was solved. Archimedes then went to the king and told him that the crown was not pure gold, but a mixture of gold with some other metal.

Using these discoveries of Archimedes, Galileo was now hoping to find a way of making some scales for weighing various mixtures of metals. He wanted these scales to be simple enough for anyone to use, accurate, and easy to carry around. Galileo succeeded in doing this by inventing the "hydrostatic balance." Unfortunately, we do not know just what Galileo's original device looked like, for details of its construction have been lost over the years.

People became interested in Galileo's hydrostatic balance. They came often to his father's shop to look at it and to ask Galileo to test various metals to see if they were pure. But few thought of buying a balance from the inventor for their own use.

Word spread in Florence, then to other cities. Vincenzo Galilei's son was quite a clever young man. He had ideas, but it was unfortunate that he was not putting them to better use! The one useful thing Galileo seemed to do well with his ideas was to explain them to others.

Here, of course, was exactly what he wanted to do! Galileo wanted to teach! He wanted to be a professor at a university and teach young men about mathematics and his scientific discoveries.

Again he turned to his friend Ostilio Ricci. "How does one go about becoming a professor?" he asked.

Ricci explained the matter to him. He would have to write letters to important noblemen, scientists, mathematicians, and politicians, telling of his work. He must try to convince them of his knowledge. He must, if possible, see them personally. Then he must ask that they try to find a position for him in a university.

Sending letters was easy enough. While some letters were ignored, he received enthusiastic responses to others. The most enthusiastic came from the Marchese Guidobaldo del Monte of Pesaro. The marchese found the hydrostatic scales interesting. He wanted to know more about the young man who had invented them. More letters were exchanged, with good results. The Marchese Guidobaldo, a nobleman of great scientific talent, promised to try to obtain for Galileo a post of mathematics professor at the University of Pisa. This would take time, however. It was very important to gain the favor of the Grand Duke of Tuscany, who could have him appointed professor.

Galileo took his hydrostatic scales, together with letters from the marchese and Ostilio Ricci, to the duke's palace several times before permission to present his petition to the nobleman was granted. The duke listened, tried not to look too bored, and promised to consider the matter.

Time passed and spring came to Florence. Galileo was still awaiting the result of his letters and interviews. In his impatience, he went to Ricci for advice. Since it might be many more months before he heard about an appointment, he would like to put his time to good use. Never had he been farther away from home than Pisa. This spring and summer he would like to visit some of the other universities—Padua, Bologna, Siena, even Rome. Could his friend help him?

Ricci encouraged Galileo. It would be well for him to travel, to meet other scientists, and to exchange ideas with them in person. Conversation, he told him, would be much more stimulating than letter writing. Ricci himself would give him letters of introduction to important scientists in several cities.

Galileo then went to his father with his plan. Vincenzo made no objections but could offer him very little money to carry it out.

Happy and free, the blood singing in his veins, Galileo took to the road. He was young and healthy. The sky was blue, the trees were bursting into leaf, the sun was warm. He walked or begged a ride whenever he could.

In this manner he traveled from city to city. At Bologna and Siena, he was received warmly by the scientists and mathematicians to whom he presented Ricci's letters. A new world opened up before Galileo; he found people had heard of him, of his pulsifier and hydrostatic scales. He spent long hours, often well into the night, discussing his ideas and listening to those of others. He made many friends. Some, like the powerful Piccolomini family of Siena, were to be of great help to him in the years to come.

At last Galileo reached Rome. There he introduced himself to the Jesuit Father Clavius, a mathematical genius teaching at the College of Rome. The older man welcomed him warmly. In a short time, a great affection sprang between the two men. They spent many days working mathematical problems together. It was with genuine regret that Galileo took leave of Father Clavius and set off for home.

In October 1587, he returned to Florence, his mind sharpened by all he had learned in these last months, his body strong and browned by the sun. But no news had come as yet about a university appointment.

For almost two years Galileo lingered on in Florence, waiting. His unhappiness at seeing no offers of a position was made even greater during that time by the sad news that his friend Ostilio Ricci had died.

Father Clavius was far away in Rome, and although Galileo wrote to the older man, it was not the same as seeing him in person. But he did have one other good friend, the Marchese Guidobaldo. This old man spared no effort to help his young friend. He even convinced the Grand Duke of Tuscany that the University of Pisa needed a professor of mathematics and that Galileo Galilei of Florence was the man for the job.

To Galileo's delight he was given a post, and in the summer of 1589, at the age of twenty-five, he returned to Pisa. Now he was not a troubled medical student who hated medicine but a professor of mathematics and science who loved his work. True, the salary was small, but somehow he would manage. No longer would he have to ask his father for money.

Galileo had returned to Pisa in triumph.

IV

Experiments from a Tower

PISA HAD NOT CHANGED IN the four years Galileo had been away. It was as if he had never left that city. The cathedral beautified the square as before, and the nearby Leaning Tower still stood at its odd angle.

This tower was Pisa's landmark. Hundreds of years before, while it was still being erected, the builders noticed that instead of standing upright, it had begun to lean over at an angle. The tower's foundation had probably not been made deep enough. Rather than tear down the half-built structure, it was decided to complete it after making sure it was steady and safe. Afterward, Pisa was known as much for its curious Leaning Tower as for its cathedral and university.

In another respect, too, Pisa had not changed. The professors at the university, despite his pulsifier, still did not like Galileo. At twenty-five, and without having graduated from college, he was again presuming to be their equal. Galileo was still finding fault with their methods and still asking too many questions. Moreover, his personality had grown more objectionable. He was too cocky, arrogantly looking down on men old enough to be his grandfather. Anyone not agreeing with his point of view Galileo considered either ignorant or unwilling to be informed.

Galileo had as much to learn at the University of Pisa as he had to teach. He had to learn to respect the knowledge of older people, even though their ideas differed from his. He had to learn, too, to be humble, to keep his temper, and to control his sharp tongue. In short, Galileo had to learn to be diplomatic.

As a teacher, the young man was not very popular. He was impatient with slower students and often passed cutting, sarcastic remarks about them. On the other hand, he spent many hours in and out of class with those who could keep up with his own galloping mind.

It was customary then in the universities for students who could afford it to ask a professor to tutor them privately for a fee. Galileo had expected to add such fees to his meager salary. Few private students sought him out, however. Unfortunately, he inspired only students who were good enough not to need tutoring. Those who might have needed his services he usually antagonized.

The young professor nevertheless surrounded himself with a small group of bright young men interested in mathematics and science. Though not much older than his students, Galileo was considered their master. He spent many an evening presenting his scientific ideas to these students.

The University of Pisa drew students from all over Europe. Among Galileo's small circle were men from England, France, Germany, and other parts of Italy. They never forgot the young man they considered their great teacher. When, in time, they returned home, they spread his teachings far and wide.

One thing Galileo could not do was to neglect his experimenting. He was searching for a complete explanation of the laws of the pendulum. He knew how pendulums behaved, and he could calculate their times of swinging, but he was not sure of the reasons why. And Galileo could not erase this "why" from his mind.

Again he read through Aristotle's works carefully for a hint to the answers, but he could find none. Aristotle stated that when an object was dropped from a height, its speed of fall depended upon its weight. The heavier the object, the faster it would fall.

Weren't his own pendulums really weights being dropped from a height? Only the strength of the string kept them from falling to the ground. Yet, if the strings were of equal length, the pendulums reached the lowest point in their swing in equal time, no matter what the weight.

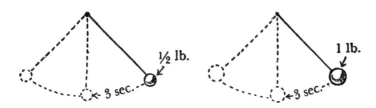

Galileo's thoughts made a great leap. Suppose Aristotle was altogether wrong? Supposing all objects when dropped from the same height fell to the ground at the same time? Only experimentation would prove this startling idea. He, Galileo, meant to find out. Again he collected objects of different weights: an hourglass, quill pen, ink, and paper. It was easier now for him to set about his task. In his earlier years as a student when experimenting with pendulums, he had learned how to work more quickly, to go right to the heart of what he was trying to prove.

Galileo climbed onto his table, held objects high over his head, and dropped them to the floor. He timed their fall with his hourglass. Always, the result was the same—equal distance, equal time!

But at best with this makeshift arrangement, Galileo could only drop objects a short distance. He needed greater heights and people to help him. One person must drop the objects from a tree or some other height, another must be below to mark their time of fall.

Galileo enlisted the help of some of his better students. Many times and from many different heights, all kinds of weights were dropped with unchanging results. As before—equal distance, equal time.

"Suppose," Galileo thought, "we dropped objects from a *very* great height. Would the effect still be the same? We shall try dropping them from the top of the Leaning Tower."

With some of his students, he repeated his experiments from the 179-foot tower. Again his conclusions proved to be correct. He was now ready to demonstrate publicly what he had proved to himself by experimentation.

A few days later, Galileo posted a notice in the university. At noon the next day, all were invited to watch him prove, with the assistance of some of his students, the law of falling bodies. The public was to assemble at the foot of the Leaning Tower for this event.

Curiosity prompted a small number of students and professors to appear there at the appointed time. Casual passersby, too, seeing a crowd, stopped to watch.

Galileo explained to those assembled what he proposed to do. He told them that, though it meant proving Aristotle was not always correct, he was about to perform a demonstration of the law of falling bodies. He was sure

The Tower of Pisa

that once they had seen his proof, they would accept the law.

His students, armed with hourglasses and cleverly constructed boxes, climbed the tower. Once again Galileo had used his skill in toy making. He had made boxes with buttons that would open at a touch, releasing objects at exactly the same time. One student remained with his boxes on the second floor, one on the third, one on the fifth, and one at the top of the tower.

At a signal, the students on the second floor released the weights from the boxes, while Galileo timed their fall. Then he signaled each in turn, and finally the one at the very top. In each case, objects dropped from the same height reached the ground at the same time.

Galileo turned to his audience, expecting congratulations to be heaped upon him. To his great surprise, the onlookers were unimpressed. Nothing this cocky, unruly-tongued, insolent youth did could impress them. Besides, he had dared to take it upon himself to disprove Aristotle! Though still in his twenties, this young man was aiming to shake the very foundations on which science had been based for hundreds of years.

Even those students and professors who did agree with Galileo's findings hesitated to speak up for him. They did not want to be known as followers of so unpopular a man. Only by their writings of his findings to mathematicians and scientists at other universities was Galileo's law of falling bodies kept from being forgotten.

About this time, another incident warned Galileo that his days at Pisa were numbered. It was the custom of the grand duke occasionally to invite scholars to his palace to attend banquets. Thus the nobleman could show that he was cultured enough to surround himself with learned men. At one of these banquets that Galileo attended, he met Giovanni de' Medici, a close relative of the ruling Grand Duke, Cosimo I.

Giovanni had invented the dredging machine to deepen the harbor of Leghorn, one of Tuscany's leading ports. He brought a model of his machine with him that evening and asked Galileo's opinion on it. Galileo examined it carefully.

In its present form, Galileo decided, it would never work. But instead of tactfully suggesting changes to improve the machine, Galileo stated bluntly that the dredger was useless.

The blow to his pride caused Giovanni de' Medici to seek revenge. He spoke against the young scientist to the grand duke. Soon Galileo noticed he was no longer being

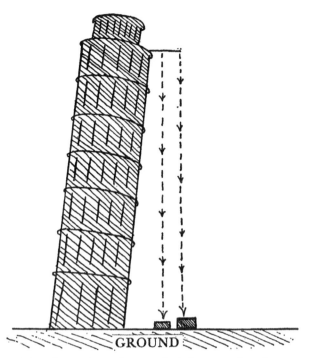
Both weights, dropped together, reached the ground at the same time.

invited to the nobleman's palace. When the invitations stopped, the authorities at the university realized he had lost favor with his powerful patron. Now their antagonism showed itself in full force.

When the term of his appointment was ended, no offer to renew it was made, nor had Galileo expected any. He was finished with Pisa. He returned to Florence in 1591, a wiser and humbler man.

V

Padua

A GREAT SADNESS FILLED THE GALILEI home in Florence, for Vincenzo had died a short time before Galileo returned. The son had absented himself from classes to come home to ease his father's last days, and to promise him on his deathbed that he would take care of the family, and pay off all the family debts.

It was quite a burden for a young man about to lose his post at Pisa. His older sister Virginia was married to a man who insisted on complete payment of a large dowry. Galileo, not daring to think from where the money would come, signed a note stating he would fulfill the obligation. The younger girl, Livia, also wanted to get married—and needed a dowry, too. Young Michelangelo was irresponsible and lighthearted. He was sure his big brother would pay for music lessons and take care of him for as long as was needed until he could find a post as court musician somewhere in Europe. And, of course, Giulia, his mother, must be supported.

Between the time of his father's death and his final leave-taking of Pisa, Galileo had not thought too much about his family burden. Now at home again, he could not ignore his duties. How could he solve his many problems?

First, he must find a new professorship. Inquiries showed that there was a vacancy at the University of Padua, near Venice, for a professor of mathematics and science. Would they, he wondered, consider him for the post after his poor reputation as a teacher in Pisa? In his need, he turned to his old friend, the Marchese Guidobaldo del Monte of Pesaro. He wrote to the marchese, asking his advice on this matter.

The marchese's answer raised Galileo's spirits considerably. The older man would be honored to help his young friend, in whose abilities he had so much faith. Galileo should first send a letter to the University of Padua asking for the position. Then he should come to Pesaro for a visit. There they would discuss how best to go about making sure of his appointment.

Galileo was delighted at the prospect of a face-to-face meeting with his friend and admirer, whom he knew only from his letters. As soon as he could, he sold his father's shop, kept just enough for himself to get to Pesaro, and used the rest for his family's needs. He promised to send them more money as soon as possible but could not say when he would return. It might be necessary to go from Pesaro to Venice to speak to the governing board of the University of Padua, the Riformatori. He hoped he might be appointed and start his work at the university immediately.

The Marchese Guidobaldo received Galileo with open arms. Although he was many years older, his admiration for the young mathematician was so great that he looked on him as an equal. He was a wealthy man, and in his comfortable home treated Galileo as an honored guest.

The two men made their plans for securing the appointment at Padua. Most important would be meeting the proper influential people, convincing them first of Galileo's ability, and then asking that they plead his cause before the three governors of the university. The marchese had many friends in Venice and Padua. When Galileo was ready to leave, he would give him letters to those who could serve him best.

Galileo wondered aloud why the old man, a good scientist and mathematician in his own right, was willing to go to so much trouble for him.

"It is not only for you," replied the marchese, "it is for the world of science that I do this. I, too, wish to enlighten men's minds. But my time has passed. My work is of little importance. You are only twenty-eight, yet you have done so much in these few years. I know that, given the opportunity, you will do much more. And what better place is there for you than the great University of Padua?"

Indeed, there would be no better place for Galileo. The University of Padua was known throughout Europe for its policy of encouraging freedom of inquiry and thought. It had thrown off the chains of the church councils at Rome and opened its doors to men of all faiths. A Protestant from Scandinavia or the Netherlands was as welcome as a devout Catholic or priest from France or Italy. The university authorities asked only that a man be willing to learn and that he respect the opinions of others.

A professor was appointed there to teach a certain subject, but no one ever told him just what aspects of that subject he should teach. Sometimes strange things happened. A book written by a professor might be banned in Venice twenty miles away, yet, at Padua, he was not prevented from teaching what was in the banned book.

Thus, an idea which might have reached a few could be spread to hundreds of men from many countries!

Galileo, for example, would never again have fear of being ridiculed for finding fault with Aristotle. There would be those who disagreed with his new theories, but they would not humiliate him. In thinking back, too, on his three years as a professor at Pisa, he saw where he had made his mistakes both with students and faculty. Those years had taught Galileo something. He believed he now knew how to make his lectures interesting and how to gain personal popularity as well.

He was eager to try his newfound strength and begged Guidobaldo to forgive him if he seemed impatient to leave for Padua with the letters. The marchese, sensing Galileo's embarrassment at his poverty, offered to lend him enough money to take care of him for a little while.

Foremost among Padua's influential men was the wealthy and learned Giovan Vincenzo Pinelli, friend of the marchese. To him Galileo brought his first letter of introduction. He was brought into Pinelli's great library, famous throughout the Venetian Republic. Pinelli read the letter, bade the young man welcome, and asked after his old friend. Then he turned his whole attention to his guest.

He told Galileo his reputation had preceded him. Pinelli knew of his pendulums, his hydrostatic balance, and his experiments with falling bodies. He would be happy to sponsor his application.

At Pinelli's house, Galileo met some of the professors of the university. To them, he was not an insolent, arrogant young man but a person who deserved respect. Under the warmth and kindness of their treatment, he blossomed forth. Any doubts he may have had about his ability to

St. Mark's Square, Venice

carry on the work at Padua disappeared. He was sure he was equal to the task.

A better picture of Venice and the University of Padua was taking shape before him. Venice, called the "Jewel of the Adriatic," was a joyous, carnival-loving city built on many small islands that were connected by bridges. Gondolas gliding along the canals provided easy means of transportation from one part of the city to another. Here, probably more than anywhere else in Europe, a wealth of printed material found its way into bookshops. It was indeed a scholar's paradise.

The city was proud to call itself a republic. Its senate was elected, as was the city's head, the Doge. The University of Padua belonged to it and was governed by its three elected members of the Riformatori. The freedom of the people and of the university was thus jealously guarded.

The Doge's Palace, Venice

Venice was also free in one other respect. Though its citizens were devout Catholics, they refused to permit the church to take a hand in any but religious matters. The authorities in Rome had no hold on the political and personal lives of the citizens of the Venetian Republic.

Galileo breathed the free air of Venice and Padua and felt like a different man. He now had friends and admirers, both young and old, eager to see him established at the university. They were glad to help him.

At last he received good news. He was appointed professor of mathematics, science, and astronomy at a salary one-and-one-half times what he had received at Pisa. At the age of twenty-eight, Galileo had reached a proud new height in his career.

Immediately he found a small house for himself in Padua, then set to work preparing his first lecture.

VI

A Master Achieves Recognition

THE RECTOR OF THE UNIVERSITY, accompanied by other members of the faculty, led Galileo into a lecture hall packed with students. He was to give his first lecture, in Latin, on the values of studying geometry.

This was a very different Galileo from the one who had first stood before his students at Pisa. It was not necessary for him to cover his insecurity by appearing overconfident. He now felt secure enough to forget about himself and to think only of his subject.

As he warmed up to his lecture, Galileo noticed he was holding the interest of the students. Soon he forgot them, too, except as people to whom he was telling the wonders of geometry.

The end of his speech was greeted with thundering applause. Students came up to him, some to ask him questions, some just to get a closer look at the new professor.

He was making good! His popularity grew by leaps and bounds. The lecture hall to which he had been assigned became too small to hold all who wished to hear him. He was transferred to a larger room, then to the largest in the entire university. Even that was packed with students wanting to hear this great professor.

Eager young men sought him out for private instruction. Many of them were from distant lands. Princes, lesser noblemen, and sons of generals competed with each other for an opportunity to study with Galileo. All the recognition he had missed at Pisa he was now getting many times over.

Realizing his rooms were cramped, he moved to a larger house, more suitable for a special purpose that he had in mind. A limited number of students were invited to live in his house for a fee and to share his table. Thus they could informally discuss all types of problems at dinner and at any odd hours they happened to be together.

To be relieved of all details of running his house, Galileo hired a married couple as housekeeper and caretaker. He seldom asked them for an account of what they spent. As long as food and drink were plentiful and the house comfortable, he was satisfied.

For the first time in his life, Galileo was earning a fair amount of money, yet he was always short of funds. Virginia's husband threatened to sue him unless the long-overdue dowry was paid. Livia had just married, and *her* husband demanded his dowry. Michelangelo made no attempt to earn a living for himself, much less help his brother by assuming some of the family responsibilities. Moreover, he wanted to come to Padua for a long visit.

Galileo had promised his father he would take care of the family, and this promise he would fulfill to the best of his ability. His housekeeper and caretaker were probably cheating him, too, but he cared little for money as long as his needs were met.

Michelangelo came to stay with his brother. Of an evening, out would come the lute Galileo still loved to play, and the two brothers would make music together. But for the most part, Michelangelo lounged around the house or loafed in the streets and wine shops of Padua. Someday, he would say, just the right position as court musician would open up for him. Finally, he announced he was going off to Poland, where he was sure he could obtain a post.

Galileo borrowed money to pay for his brother's trip and to satisfy his brothers-in-law's demands. This done, he was able to turn to his work with a clear mind.

Galileo's house was an interesting one. Because his lodgers came from different countries, they had no common language but Latin. This, then, became the language of the dinner table and of discussions that often lasted far into the night.

At the university, Galileo lectured almost entirely on mathematics and astronomy. At home, he explained his theories and the results of his experiments in physics. He kept many notes of his experiments but so far had not put them together in the form of a book. That was a task for the future. For the present, information about his experiments was spread through Europe mostly by his students.

Since many of his private students came from noble families, Galileo discovered that one of their main interests was the study of fortifications. Local wars were frequent between one duchy and another, and it was necessary that heirs to ducal crowns be well acquainted with military science.

Though Galileo knew little about fortifications, he felt it very important to become an expert on the subject. For one thing, private pupils paid very well for such instruction, and he was constantly in need of money to supply his family's demands. For another, it was well to have as many influential friends as possible throughout Europe. But most of all, his mind refused to be idle. If there was something he did not know, it was a challenge. He had to learn more about it.

Thus, the professor of mathematics set out to teach himself the science of fortifications. It was not hard at all! He was able to combine his knowledge of mathematics and his experiments in mechanics with what was known about fortifications. Before long, he mastered what he had set out to learn—and better yet, improved on it.

Galileo showed his students how it was possible to measure forces mathematically, how to calculate the force of a weapon, and how to plan defenses with great exactness. He taught them how to calculate the path a cannonball would take so that they could place a cannon to the best advantage. He even thought of a device, an improvement on the compass used in drawing, that would make such plans and calculations easier to prepare. Someday, he told himself, he might really develop this compass and sell it.

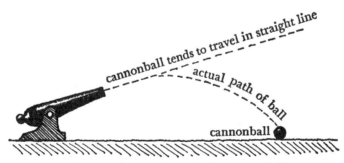

Gravity changes the path of a cannonball

Very likely, much of his interest in fortifications lay in his desire to make use of the science of mechanics in which he was experimenting. His lively imagination was always trying to find some practical use for his scientific findings.

He had used his laws of the pendulum to make a pulsifier. After reading and studying Archimedes' laws of objects in water, he devised a use for this information in the hydrostatic balance.

Always he would think: "To what practical use can I put these facts? Just knowing them is not enough."

Galileo's fame as a teacher of fortifications spread. Princes were honored to lodge in his house and receive instruction from him. He was a happy man. Life was good to him. He was doing work that he liked and that was useful; he earned a decent living, and he had many friends.

For two years fortune smiled on Galileo. Then disaster struck. He caught a chill and fell ill. What at first seemed to be nothing but a bad cold became serious. His whole body was racked with pain. Friends, realizing this was no longer a simple cold, called in Dr. Fabrizio, the greatest doctor in Padua.

Dr. Fabrizio examined him carefully. The pain was worst in his joints, especially in his left wrist and knee. Regretfully, the doctor made his diagnosis known. Galileo was suffering from arthritis. There was no known cure for this disease, only medicines to relieve some of the pain. With fortune, the professor would get better again, but he would never be completely free of the possibility of future arthritic attacks. When he recovered from this one, years might pass before another came, but come it most likely would.

The sad news spread quickly. Friends came to care for him, to visit him, and bring him books to help pass the time in reading when he was alone. Strange and interesting books reached his bedside, some of which he might never have had the time to read otherwise. One of them was by a Polish monk, a mathematician-astronomer, Nicolaus Copernicus.

A MASTER ACHIEVES RECOGNITION

A strange book was this indeed! Published in 1543, it was still, in 1594, not too well known, though some of its revolutionary ideas were occasionally discussed among astronomers. Some even thought the ideas in the book made sense. Copernicus stated—and set out to prove—that the sun stood still, and that the earth moved about the sun! Heretofore, of course, astronomers throughout Europe taught the opposite—that the earth stood still, and the sun moved around *it!*

This whole business was worth looking into thoroughly. Could Copernicus be right and all other astronomers wrong? Galileo promised himself that as soon as he was well again, he would try to find the time to explore the great possibilities Copernicus presented.

Slowly his strength came back to him. He was able to resume his duties at the university and to teach his private pupils. But for the rest of his life, he knew he would have to watch his health carefully.

VII

Golden Years

WITH HIS RETURN TO HEALTH, life again went smoothly for Galileo. He taught his classes at the university, worked with his student-lodgers at home, and experimented constantly. Life without experimentation was unthinkable. Others might accept a statement made by a scientist without questioning it, but Galileo demanded proof. Even proof written about by another man was not enough. He maintained that if a scientific statement were to be accepted as fact, then anyone with the same materials and the same methods should be able to obtain the same results.

We accept this scientific method today as the only proper one. In fact, we can imagine no other. Yet Galileo struggled with the scientists of his time to establish this method of inquiry. Because of his efforts, he has earned the title of "Founder of Experimental Science."

Little by little, he was perfecting the proportional compass. He was almost ready to put the final touches to it. Just one or two more adjustments were necessary. But, the unexpected happened. Galileo fell in love.

On one of his trips to Venice, he met Marina Gamba. Her beauty attracted him at first, then her gay, light-hearted chatter. He needed her to make his life complete. But he knew that to marry Marina and to bring her into his home in Padua, to a house filled with students and everlasting scientific discussions, was unwise. She was not at all interested in his work and would soon become bored and unhappy.

Much as he loved Marina, he loved his work even more. Nothing must ever interfere with it. Marina's constant presence in his house could only distract him and his students. He could ask the students to leave, marry Marina, and live a quiet, domestic life, or forget her.

But perhaps she would consent to a third solution. Marina might be willing to have him furnish a house for her near his own in Padua. Then he could spend time with her whenever possible. She, on the other hand, would not be burdened with the work of keeping house for a lot of noisy, wine-drinking, careless young men. Most of her time would be her own. She would live like a lady, in her own house, and have no financial worries.

To Galileo's great joy, Marina agreed to this arrangement. She understood nothing of his work. In fact, she was a little afraid of appearing dull-witted in the clever company Galileo kept. Living in her own house and having Galileo, too, was what she wanted.

Thus Galileo bought and furnished a home for Marina near his own and brought her there as his common-law partner.

Days at the university, evenings with his students and friends, some time with Marina—all these made his life full. To add to his happiness, he learned that his university appointment would be renewed when the time came, and his salary would be increased. He was now able to turn his attention finally to perfecting the proportional compass.

The proportional compass

It was an interesting device. Though ordinary drafting compasses had been in use for a great many years, Galileo's improvements increased the uses to which it could be put. His could be used, among other things, to enlarge small maps accurately and easily, as well as to simplify very difficult mathematical calculations. Today we call it a *sector*, and draftsmen still use it, with some slight changes in Galileo's original design.

Once a person learned to use the proportional compass, or, as Galileo put it, the "geometrical and military compass," it was simple, but learning how proved complicated. Therefore, Galileo wrote out carefully, by hand, instructions for its use. Students became interested in it, and some offered to buy the compasses at a good price. Galileo went into the compass-making business in a small way.

He hired one man to make the compass to his order, and another to write out, in Italian, instructions for its use. Though Latin was the language used in the university and for the most part with his students, Galileo preferred having these instructions written in the language of the ordinary man. He wanted his compass to be used by as many people as possible, not just by scholars. To enable craftsmen, who knew no Latin, to learn its operation, the instructions had to be in Italian.

Business flourished. Soon his compasses were more and more in demand, and he was earning a good income from their sale.

At about this time, Galileo invented another interesting device: the air thermometer. This device was so simple, it was surprising that no one had thought of it sooner.

It consisted of a long glass tube, open at one end, and widened into a bulb at the other. When the tube was filled with colored water and inverted into a pan of water so that the open end of the tube was below the surface, a trapped air space would be left in the bulb. The thermometer was now ready for use.

Heating the bulb the slightest bit, even with the hand, caused the air in its top to become warm and expand. The expanding air, needing more space, would then push some of the water down the tube and out into the pan. The warmer the air in the bulb, the more water was pushed out. When the heat was removed and the air in the bulb cooled, it contracted and left some empty space in the tube. Water from the pan rushed in to fill this empty space.

If the tube were marked with a scale to measure the height of

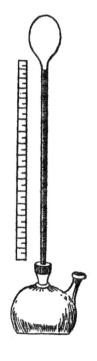

Galileo's air thermometer

the water, this scale could be used to measure the temperature of the object heating the air in the bulb. Thus we have a kind of thermometer.

Galileo knew his thermometer was not entirely accurate or perfect in many respects. He could not measure very cold things because the water in the tube would freeze. Then, too, he noticed that on different days, the height of the water in the tube would be slightly different, even though no heat was applied to the bulb. Particularly on cloudy or rainy days, the water in the tube was lower, even though on those days the air outside was warmer than on clear days.

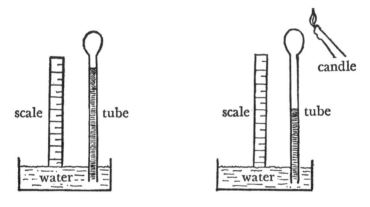

How an air thermometer works

He did nothing else about his air thermometer. In 1643, after Galileo's death, his pupil, Evangelista Torricelli, experimented further with it, following suggestions made to him by Galileo in the last few months of his life.

It was likely, the master had said, the air pressure varied with the weather. This pressure of outside air on the water in the pan probably affected the height of the water in the tube. Torricelli, in his experiments with this air pressure, substituted mercury for the water in both tube and pan, and finally invented the mercury barometer, which is still in use in forecasting weather.

Others set to work improving the thermometer. The liquid was sealed in the tube so that outside air pressure would not affect it. Later, as water in the tube proved to be unsatisfactory, alcohol or mercury was used. In 1714 a German scientist, Gabriel Daniel Fahrenheit, constructed an accurate scale for measuring the rise and fall of the liquid in the tube, and thus the temperature of the object causing the change.

The thermometer now looked very different from Galileo's early air thermometer, but it had its beginnings in his pioneer invention. To him must be given the credit of thinking of two things. One was measuring temperature by the expansion and contraction of a material. The other was the use of a glass tube and a liquid in this process.

All was going very well indeed for Galileo. Though he had several attacks of arthritis, none were severe. In 1600, he had further cause for happiness. A daughter was born to him and Marina. They called the child Virginia, after his oldest sister. Another daughter, born in 1601, they called Livia, after his younger sister.

Galileo spent many an evening with his two children. Marina asked very little of him. She was content to remain in her house and bring up the children. A simple woman, she still made no attempt to understand any of Galileo's work. All her interests were centered on the house and the children. She knew their father was considered a great man, and she stood in awe of his brilliant mind.

Again the Riformatori renewed his appointment at the university for another six years, with an increase in salary. They were pleased to have him at Padua. But Galileo had never forgotten that he was a Florentine, though recognition had not come to him in his own city. He longed for the place of his childhood and dreamed that perhaps someday he would be able to go back there to live and work.

To help make this dream come true, he formed a long-range plan to keep his name before the ruling family of Tuscany, of which Florence was the chief city. The Crown Prince Cosimo was a young boy who, in time, undoubtedly, would need a tutor in mathematics and science. In 1601, Galileo wrote to the grand duke in Florence and asked that he be permitted to serve in that capacity. He hoped thus to be kept in mind when the proper time came.

In 1605, Galileo was invited to come to the summer palace of the Medicis, rulers of Florence. There he would instruct young Prince Cosimo during the summer vacation.

Delighted with this opportunity, Galileo left Padua for a six-week stay at the Medici palace. He found the fifteen-year-old Cosimo—a serious, intelligent youth—eager to learn all that his tutor offered. The young prince and the middle-aged man became fast friends. When the time came for him to return to Padua, Galileo knew he had left behind in Florence an admirer and protector.

Sales of his compasses were booming now. To control their manufacture, Galileo set up a small shop in the rear of his house where he could supervise the workmen. It was becoming impractical, too, to continue writing out the instructions by hand. Galileo set himself the task of writing clearly and in detail, in the form of a booklet, just how to use his proportional compass. This he dedicated to the Crown Prince Cosimo de' Medici. Then, in 1606, he set up a small printing shop in his house where these booklets were printed.

Now he was able to supply many more compasses. Orders came in from all over Europe. Business could not be better.

That year, too, Marina bore him a son, whom he called Vincenzo, for his father. Now he had a small family of his own. But though he came often to spend time with his children, Marina became more and more indifferent to him. As he became more famous, she grew more withdrawn. He saw that someday they might part.

One day, in 1607, a student came to him excitedly waving a paper. It was a Latin explanation of the use of his compass, translated almost word for word from his Italian. Galileo was angry. Careful investigation brought out that the translation had been made by Baldassar Capra, a student at the university. Capra had cleverly gotten Galileo to teach him the use of the compass. Then he went about making some on his own, then selling them elsewhere together with the Latin booklet explaining their use. Moreover, he claimed that Galileo had stolen the idea from *him!*

This last angered Galileo even more. He brought charges against Capra before the university officials. After looking into the matter, the officials decided in favor of Galileo. They ordered Capra to refrain from making or selling the compasses. They also insisted that he take back those he had already sold and refund the money.

Capra could not remember all to whom he had sold the compasses. In self-defense, in order to clear his own name, Galileo decided to write and publish a short book saying he was the inventor of the geometric and military compass, and that any other claims were false. This book was distributed throughout Europe, wherever it was at all possible that Capra had had customers.

This book was clever and witty. Galileo read it over again and was so pleased with it, he suddenly realized he had another hidden skill. He could also write well. He need have no fear of his ability to put words on paper. He would not wait much longer to write of his discoveries for all the world to read.

VIII

A Simple Spyglass Becomes a Telescope

Now that Galileo knew he could write clearly and well, he became more and more eager to get to work on his books. He had plans for a number of them. One was to be about his new science of moving objects, three were to be on mechanics, and several would be shorter ones on such topics as sound, vision, color, and tides.

While he went about his other duties, he kept thinking of the books but could find no time to start writing. Teaching at the university, tutoring his private pupils, and supervising the manufacture and sale of his compasses filled his days. He resolved to find some way to gain time for his new projects.

If only he did not have to teach his classes at the university! He loved working with his private pupils, but they were a noisy lot, demanding much of his time and energy. Galileo wished he could give up his pupil lodgers, yet he needed the money they paid him.

Besides sending money to his family in Florence, he had Marina and three children to support. Michelangelo, too, never seemed to be able to manage for himself. He was married now and had children of his own, yet he still frequently asked his brother for help.

Someday, somehow, Galileo must find time for his writing, but it would hardly be likely as long as he remained at Padua. There was one big disadvantage in working for the Republic of Venice. Because it was a republic, the Riformatori of the university were responsible to the people, and the people insisted that a man perform the work for which he was being paid. If he was hired to teach, he should teach and not devote all of his time to experimenting and writing.

A ruler like a grand duke, however, was responsible to no one. If Galileo could convince such a one of the value of his work, he might find a patron. Then, except for a few small tasks, he would be free to pursue his own work. Of course, he would dedicate his books to the patron, and thus the nobleman would become famous, too.

Word reached Padua in February of 1609 that the old Grand Duke of Tuscany had died and that Galileo's one-time pupil who had liked him so well, Cosimo, had become ruler in Florence. Here was Galileo's chance. He wrote to Cosimo's Secretary of State, Belisario Vinta, asking that he be kept in mind for the position of court mathematician, which his old friend Ostilio Ricci had once held.

Vinta promised he would not forget. At present, however, too many other matters of greater importance occupied the nineteen-year-old duke.

Galileo, though impatient, knew that he could not rush matters. He was confident that sooner or later his young friend, Grand Duke Cosimo II of Tuscany, would invite him to become his court mathematician. In the meantime, things were moving too quickly in the world of science for Galileo to sit by idly waiting.

Johannes Kepler, a great German astronomer, wrote a book that was published the same year, 1609. He called it the *New Astronomy*. Galileo obtained a copy of the book and read it with great interest.

Though he had never met Kepler personally, he considered him a friend. For many years the two men had been corresponding with each other, exchanging views and information on many questions in astronomy. Kepler had stated that the more he looked into it, the more he was coming around to Copernicus' theory of the solar system. Galileo had admitted that he, too, was reaching the same conclusion but needed more proof before he could accept such a revolutionary theory wholeheartedly.

Kepler's *New Astronomy* was a long book, and for the most part quite dull. Few bothered to study it. The astronomer went into great detail describing all the research work he had done. He included long accounts of all the wrong leads he had followed and all the blind alleys which he had gone down. Buried among these accounts was a statement that Copernicus' theory was correct. The earth and planets moved about the sun in regular paths that could be calculated.

Galileo waded through the faulty Latin of the badly written book. Once he unearthed the real information the book had to offer, he saw that Kepler had worked out a proof of Copernicus' theory to the German astronomer's own satisfaction. Galileo, however, was not completely convinced. He hoped to find proof of it someday. When he did, he would write a book in simple language, probably in Italian. Then any reasonably well-educated person could read it, not just a few Latin scholars.

The possibility of finding this proof was nearer than Galileo suspected. By means of a tube and two pieces of glass ground exactly right, he was to reveal the glory of the heavens. In doing so, he was to reach a great turning point in his own life.

June of 1609 began like any other June, with the year's work at the university drawing to a close. One day a rumor reached Galileo. An eyeglass maker in the Netherlands, Hans Lippershey, had invented a curious tube that made things appear larger if you looked at them through it. Like any other rumor, some said it was true; others denied it.

A few days later a letter reached Galileo from a former pupil who had returned to his native Paris. This French nobleman confirmed the report. He wrote that though he did not know how Lippershey had done it, the lens maker had indeed constructed a spyglass that made distant objects appear larger.

A spyglass that made things seem larger! Such an instrument had a great many possibilities, both on sea and on land. Galileo wanted to construct one, too. Classes at the university were over for the summer, and he had the time to spare for this interesting project. With no further information about Lippershey's invention, he set to work that same evening.

He read everything he could find about lenses. There was so little, however, that he finished in a few hours. Then, with pen and paper, he began to draw figures of lenses and tubes, and make mathematical calculations. Late into the night he worked. Before morning, he thought he would be able to make a spyglass of his own.

Galileo knew that we are able to see an object when light rays from it strike our eyes. These light rays travel to the eye in a straight line unless something interferes with them. Lenses can bend these rays from their straight path. A convex lens, one which is thicker in the middle than at its edges, bends the rays inward. On the other hand, a concave lens, one which is thinner in the middle than at its edges, bends light rays outward.

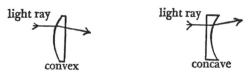

Placed at the proper distance from each other, the lenses could enlarge the appearance of an object.

Here is how his spyglass would work:

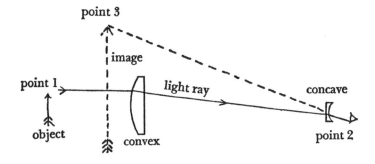

Rays of light from an object (point 1) would first enter the convex lens, and so be bent inward. When these rays reached the concave lens, their paths would be changed so that they would be bent outward, to point 2. An eye placed at point 2, being able to see only in a straight line, would see point 3. In the same way, light rays from all other points of the object viewed would be received, thus producing an enlarged image.

Galileo would also need a tube into which he could set the lenses to keep them the proper distance apart. The tube would have to be adjustable for viewing objects at different distances and to allow for differences in the vision of people looking through it. He could accomplish this by using two hollow tubes, one of which slid into the other.

The next day he began experimenting. He spent many hours patiently grinding and polishing pieces of glass until they were exactly as he wanted them. Always, he made them in pairs, one convex, one concave.

At last, the lenses were ground exactly right. He prepared a double sliding tube of lead, fitted a large convex lens in one end and a small concave lens in the other. With his eye at the concave lens, he pointed the tube at a tree outside his window and adjusted it. Galileo was amazed! The leaves looked almost as though he were holding them in his hand! He could see the veins in the leaves, even the spots eaten away by insects!

Then he turned his spyglass toward the bell tower of the church. That, too, appeared nearer and larger. He had succeeded in making a spyglass through which, in his own words, he "perceived objects satisfactorily large and near, for they appeared three times closer and nine times larger than when seen with the naked eye alone."

Galileo was not satisfied. He knew that, with patience and work, he could improve the spyglass to make things appear larger and closer. All during the summer of 1609, he worked on his spyglass, calculating, grinding lenses more accurately, and changing the length and width of the tube. His next spyglass enlarged objects more than sixty times.

Close friends who visited Galileo that summer noticed his complete concentration on his new interest. Curiosity made them ask to be permitted to look through his spyglass. They were astounded at how large and how close distant objects appeared, and spoke of practical uses for his invention. One suggested using it to watch ships offshore so that help could be sent if necessary. Another thought it might be used to watch an approaching enemy in wartime.

There was some talk, too, of giving the instrument a name other than "spyglass," but no one could find a word that would describe it better. Not until two years later, in 1611, was it given the name "telescope." It was a new word, composed of two Greek words meaning "see" and "far." The telescope was indeed an instrument that was able to see things from afar.

Meanwhile, all agreed it would be gracious of Galileo to present a spyglass as a gift to the Doge of Venice. These seventeen years the Venetian Republic had treated him well, and the Doge, as its chief representative, would appreciate this gesture.

Galileo felt this idea was good. He made a handsomely decorated spyglass and, late in August 1609, presented it to the Doge. The old man was both touched with the gift and impressed with its usefulness. He showed it to the members of the senate and asked that further recognition be given Galileo for his services to the republic.

Within five days Galileo received a message from the Riformatori that he had been appointed for life to his position at the University of Padua and that his salary was to be doubled. His friends congratulated him on his stroke of good fortune. Galileo was secure for life.

IX

The Heavens Reveal Their Splendors

WITH THE COMING OF AUTUMN, classes at the university started once more. Gone were the free days and nights of the summer. Now it was necessary for Galileo to steal time from his regular duties to continue improving his spyglass.

Though he had impressed many people with this instrument, Galileo was still not content. He was convinced he could make a much more powerful one, and the desire to do so was uppermost in his mind. Performance of his daily tasks became a chore to be finished as quickly as possible. Only then did his real work begin.

Curves and angles danced before his eyes. Figures and calculations haunted his dreams. He *must* find just the proper curve of lens and length of tube to magnify objects a thousandfold.

Even the arthritic pains, which so frequently came with the cold, damp weather, did not deter him. Marina and the children seldom saw him. Galileo lived just for the hours he spent working on his new spyglass. He spared himself neither labor nor expense in making a number of spyglasses, each one better than the one before.

Meanwhile, Marina announced plans of her own. She was tired of a great man who spoke of little except scientific ideas that were meaningless to her. Being a simple woman, she wanted a simple husband and an ordinary home life. A friend she had known in Venice before coming to Padua offered her that kind of life. She intended to leave Galileo and marry this man.

Galileo's one thought was for the children. Marina could go her way if she wished. She had told him some time ago that if the appointment as court mathematician would come, he would have to go to Florence alone. She would never be happy in those strange surroundings among such grand people.

The two reached a friendly agreement. Their daughters, Virginia and Livia, were to live with Galileo's family in Florence. The son, Vincenzo, was to stay with his mother until he was older, and Galileo would give Marina and her new husband money for his care. When the boy grew older, he would join his sisters in Florence. So the entire matter was settled peacefully.

Now there were no personal problems to interrupt the astronomer's evenings. Like a child with a new toy, he examined each part of his latest spyglass, waiting for a clear night. He had already used his invention to examine everything he could of the surrounding countryside. The starry sky awaited him.

Late in the autumn of 1609, Galileo turned his spyglass toward the heavens. Such wonders met his gaze that even he had not imagined! The velvety dark sky was studded with countless stars, some large, some small, some that he had seen before, some entirely new to him.

The bright constellation Orion shone in the sky. With the naked eye, the three bright stars of the belt and the six in the sword were a familiar sight in the autumn and winter. But as he looked at Orion through his spyglass, Galileo counted, in addition to those nine, eighty smaller stars near the belt and sword which had been invisible until now.

He explored the six stars of the Pleiades. Though they were called the "Seven Sisters," the seventh star was so faint it was hardly ever visible to the naked eye. And now, near, also occupying the little space in the sky in which they appeared, he saw forty other stars!

Carefully and accurately, he drew diagrams and made notes of what his spyglass revealed. Galileo firmly believed that all of his previous inventions and discoveries were unimportant compared to this. Very soon he must tell the world what was revealed to him. But before that, he wanted to examine the crescent moon that had just appeared in the sky.

All of his life Galileo had thought the moon to be a smooth heavenly body shining by its own inner light. He taught his pupils that the moon's curves were as perfect as though they were drawn by a compass. Imagine his astonishment to observe through his lenses that the inner curve of the crescent was as irregular as when he looked at a distant mountain range!

Night after night he watched the moon through the spyglass. Neither cold nor his arthritic pains kept him from his observations. Each night as the moon grew rounder and rounder, he saw more and more astonishing things on its surface.

In making comparisons with what he saw on Earth, Galileo became convinced that there were indeed mountain ranges on the moon. These appeared much more jagged than those on Earth, and much taller, at a height he estimated to be about four miles. He watched bright and dark spots on the moon's surface and noticed how some of them changed as the moon rose higher and higher in the sky. The shifting of the light and dark reminded him vaguely of something familiar.

Of course! Shadows on mountains and valleys when the sun cast its light on them! When the sun rose in the morning, the tops of the mountains first received its light. Then the mountainside was bathed in light, but the side facing away and the valleys below were in shadow. As the sun climbed higher in the sky, the shadows shifted. At noon, the valleys received full light. Gradually, as the day wore on, the sun lit up different sides of the mountain, and the valleys again became shadowed.

Was it possible? Was the moon, like our earth, receiving its light from the sun? What, then, of all the astronomers' statements that the moon was a perfect body, shining by its own light?

Galileo looked again at the now full moon. Some of the larger, darker areas were not affected by the shadows. It was as if someone were looking down at the sea from a mountain. Viewed from a height, the sea appeared darker in sunlight than the nearby land. These, then, must be seas too, and he made up names for many of them.

Whenever the sky was clear, Galileo made his observations. He drew maps of the surface of the moon, almost as if he were drawing a map of the earth. He worked at a feverish pace, wanting to finish before the moon waned back into its old crescent.

When the young crescent moon again appeared in the sky in the early evening, the astronomer looked at it with his naked eye to admire its beauty. Something he had

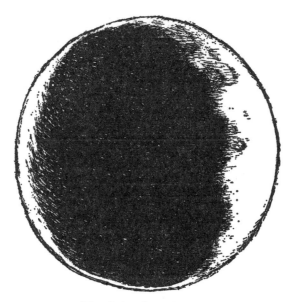

"Earthshine" on the moon

noticed before but never understood attracted his attention. Past the horns of the bright crescent, the rest of the circle of the moon was outlined by a very faint light.

He turned his spyglass again toward the moon. This time he was not looking for mountains or valleys but concentrated on trying to find a hint as to the cause of the glow. Through the lenses he examined the part of the moon that faced away from the sun. If the theory forming in his mind was correct, that part should be completely black. Instead, it was very, very faintly lit, as though it were receiving light from yet another heavenly body that was not too bright.

Galileo paused in his thoughts. He considered carefully those heavenly objects whose light might affect the moon. Suddenly something struck him, and he took pen and paper and began to calculate.

With his new calculations and what he already knew about astronomy, he reached a conclusion. At the time of the new moon, which was visible only in the late afternoon and early evening, a large part of the sunlit side of the earth was facing the moon. Just as the moon sent us light which *it* received from the sun, we sent the *moon* light which *we* received from the sun.

The faint glow of light on the dark surface of the new moon was therefore really *earthshine*. As the moon's position changed each night, it was farther and farther from the lighted side of the earth and received less and less earthshine. When it was full, it received none.

By this time, though most other astronomers thought differently, Galileo was sure the moon had no light of its own, but shone by the light it received from the sun, and which, in turn, it sent back to the earth. He decided to leave his observations of the moon for a while and examine the stars a little longer. Then he would include in one book all of his observations and theories regarding the heavens.

The wonders he saw made him eager to improve his spyglass further so that he could see even more. On cloudy or rainy nights he worked with his tubes and lenses. Early in the winter he succeeded in making so fine an instrument that objects seen through it appeared almost a thousand times larger and thirty times nearer than when seen with the naked eye.

That band of hazy light stretching across the sky on starry nights, the Milky Way, caught his interest. He observed through his new tube and lenses that the Milky Way was not a band of clouds, but thousands upon thousands of stars grouped together in clusters. Wherever he pointed his spyglass, a vast crowd of stars came into view.

Next he directed his instrument to other faint, cloud-like patches in the night sky. Because they looked misty to the naked eye, they were called "nebulae," from a Latin

Galileo's spyglass

word meaning "mist." In particular, he examined the nebula near the constellation Orion.

Here, too, the "mist" disappeared when viewed through his glass. The nebula became a cluster of many individual stars, some bright, some faint. Although each star separately escapes our sight because of its distance or size, Galileo reasoned, together they form clusters whose light, because it comes from so far, appears hazy to our naked eye.

On the night of January 7, 1610, Galileo pointed his latest spyglass at a part of the sky near the planet Jupiter. He noticed three small stars in a line near the planet, two to the left of it, one to the right, placed like this:

He remembered that just the night before, the three stars had been in a line, but all three had been to the *left* of Jupiter.

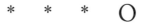

Why should a star change its place? Jupiter could be expected to move out of position since it was a planet, or "wandering star." There were all together five such wandering stars—Mercury, Venus, Mars, Jupiter, and Saturn, all visible to the naked eye. No one had ever seen any others. Planets change their position in the sky in such a regular way that their paths were well known by astronomers. Could any of these three tiny objects be new planets?

Galileo decided he must know. Every night he would watch them to see what happened. On January 8th, he was astonished to find a different arrangement of the three stars. Now all were to the right of Jupiter!

He wondered greatly about this. According to the known path of Jupiter, it should not be traveling to the left, yet there it was! Perhaps Jupiter was not traveling to the left, but the small stars were moving toward the right.

Two nights later only two stars appeared, both at Jupiter's left, still in a straight line.

Where was the third? It might, of course, be hidden behind Jupiter. Then this meant that the "stars" were traveling *around* the planet. He hesitated to call them stars any longer, for these wandered as no stars ever did.

Each night he noticed a slight change in their position. On the night of January 13th, he saw *four* stars around Jupiter, placed like this:

* O * *
    *

He watched them a few nights longer. There was no doubt now. The four were really satellites going around Jupiter like moons. As Jupiter traveled across the sky, they traveled along with it. One might almost say they were a part of Jupiter itself.

What wonders were being revealed to Galileo nightly! There was still so much to learn and so little time! Wakeful nights watching the sky and days filled with necessary work would soon wreck his health completely. He must give up one or the other.

Cosimo II, Grand Duke of Tuscany, had still done nothing about his appointment as court mathematician. To prod him, and also because he really liked the duke, Galileo hit upon a twofold plan. First, what better name for Jupiter's four new "stars" than one that would honor the duke's family? He called them the four *Medicean* stars.

Second, he finally began to write his book announcing to the world the discoveries he had made by means of his spyglass. The book was written in Latin, in clear, simple language. Galileo wanted astronomers and scientists throughout Europe to get the information he had to give as quickly as possible. Since Latin was the only language common to all of them, they could read the book without waiting for translations.

In March 1610, the book was published in Venice. Galileo called it *Sidereus Nuncius*, or *The Messenger of the Stars*. Before going into a description of what he had observed, the author wrote a dedication:

"To the Most Serene Cosimo II de' Medici, Fourth Grand Duke of Tuscany."

In the dedication he also wrote that, though many great men were honored by having statues of them made, or buildings named after them, he would honor the grand duke by naming the four stars for him. Buildings and statues are destroyed by time, but the stars remain in the sky forever.

Then he sent a copy of the book and a spyglass to the duke in Florence and awaited the results.

X

A Native Son Returns

THE MESSENGER OF THE STARS was an immediate success. News of Galileo's spyglass, and of what he saw through it, had already spread by word of mouth. Now those who knew Latin wanted to read for themselves, in the astronomer's own words, what he could tell them.

Five hundred copies of the book were printed at first. As soon as they reached the bookshops, eager readers snatched them up. Though travel and communication were slow in 1610, almost every mail pouch arriving in Padua brought a request for a copy of the book. Couriers on horseback came bearing letters asking for *The Messenger of the Stars*. Roads became muddy with the thawing of winter snow, wagons and coaches bogged down in the mire, but still requests for the book came.

Within three months, Galileo received orders for five hundred more copies from all parts of Europe. His printer in Venice went to work setting up the presses to print them. From Germany, Johannes Kepler, the great astronomer, wrote praising the book and asking that he be given permission to print an edition of it in the city of Frankfurt.

The Messenger of the Stars was on its way to being a best seller of its time.

Together with orders for his book came demands for his spyglasses. Galileo sold them as fast as he made them. He dared not leave the grinding of the lenses to assistants, for he wanted to be sure each lens was made as accurately as possible. There were not enough hours in his day for all the work piling up on him.

The professor of mathematics and astronomy at the University of Padua had rapidly become world-famous. People wanted to see him personally and hear him speak of his observations. The whole university turned out to listen to his public lectures.

Some scoffed at him and said he was performing tricks with his lenses. When invited to do so, others refused even to look through his spyglass. Still others, timid people, closed their ears to anything that Galileo said. They had always believed the moon to be perfectly smooth and the Milky Way a special kind of "night cloud." To permit themselves now to think differently, they believed, was unwise. Fear of being laughed at, or worse, of being called "radicals," made such people refuse to listen to any of Galileo's new ideas about astronomy.

However, many of the people who came to his lectures went away convinced that Galileo had indeed opened up a whole new world in the study of the heavens. They were proud to call themselves "Galileists," followers of Galileo.

When the Easter vacation came, Galileo took time out from his lectures and lens-making to pay a visit to Florence. He longed to see his daughters and to speak with Belisario Vinta, Grand Duke Cosimo's secretary of state. Thus, he hoped, by a personal appearance, to hasten his appointment as court mathematician.

Virginia and Livia were overjoyed at seeing their father. He spoke to them of the events of the last few months in Padua, and they told him of their life in Florence. Neither of the girls was very happy living with the scolding grandmother or the noisy cousins. They wished they could go to some convent school. The older girl, Virginia, though only ten years old, had already decided that she wanted to become a nun, and Livia, too, thought she would join her sister.

Galileo asked their aunt if she knew of a suitable convent school. She told him of the Convent of San Matteo at nearby Arcetri, where the girls would be well cared for by the kindly Mother Superior. When they were older, if they so wished, they could join the convent as nuns.

Relieved that his daughters were happy with this new arrangement, Galileo turned to his other business. He visited friends, wrote letters to the grand duke, and finally made an appointment to see Belisario Vinta.

The secretary of state was a busy man and could spare little time for the astronomer just then. He suggested that instead of spending many hours discussing the terms of employment at the court, Galileo should write him a letter stating exactly what he wanted. He, Vinta, would then take the matter up with the grand duke himself. Thus Galileo could be assured that his appointment would be forthcoming in the very near future.

Following his return to Padua, Galileo wrote a long letter to Vinta setting down the terms of employment at the Court of Tuscany.

"I am determined," he wrote, "to settle my future career and devote all my attention to bringing to fruition all my labors and studies of the past, from which I hope to win some fame.

"Here," he continued, "I have a perfectly secure salary for life, and can earn much more from private instruction as long as I go on teaching gentlemen from abroad, and by taking scholars into my house. But because giving private lessons and taking scholars as boarders constitutes something of an obstacle to me and impedes my studies, I should like to be completely free of these. Hence if I am to return to my native land, I desire that the primary intention of His Highness shall be to give me leave and leisure to draw my works to a conclusion, and without being occupied in teaching."

He went on to explain that he would not really be deserting the teaching profession. By writing books on his science, he would reach many more pupils than those in his classes. The salary he asked was the same as he was receiving at the University of Padua. He would also expect to have a private income from the sale of his books and inventions.

The letter was sent off in May. While awaiting a reply, Galileo turned his spyglass to the planet Saturn. Its appearance, when magnified, was different from any of the others. He could not quite understand what gave it this shape:

The lenses he had were not powerful enough to give him a clear image of the planet. Mistakenly, he thought Saturn had moons attached to it. Years later, in 1656, the great Dutch astronomer, Christian Huygens, recognized these "moons" as Saturn's rings.

July of 1610 came, and with it good news. Grand Duke Cosimo II of Tuscany was pleased to appoint Galileo Galilei to the post of court mathematician and philosopher, and also a professor of mathematics at the University of Pisa. The professorship was a title of honor and carried no duties with it. His salary would be the same as at Padua, and he would be free to carry on his research and writing. All that was required of him was to teach mathematics and science to any children the grand duke might have. Occasionally he would be asked to attend dinners at the ducal palace and discuss scientific matters with the grand duke and his guests. His duties would start in September.

At first, the Doge, the Venetian senate, and the Riformatori of the University of Padua were angry at what they took to be Galileo's ingratitude to them. They had taken him in when he was comparatively unknown, had treated him well, and had heaped honors upon him. At Padua he was secure for life. He could not ask for more.

Patiently, Galileo explained to them his reason for seeking the change. They understood and forgave him, for they could not offer him the complete freedom for his research that he wanted.

Summer passed quickly. Much had to be done before Galileo took leave of Padua and the Republic of Venice. Arrangements for the printing of his books and the manufacture of his proportional compass were made, and his house was put up for sale.

In his eighteen years at Padua, he had made many fast friendships. Regretfully he bade his friends farewell, promising to return for a visit. His closest friend, Giovan Francesco Sagredo, however, was absent from Venice on a mission for the republic. He had been gone for almost a year. Months later, when Galileo was already settled in Florence, a letter from Sagredo reached him.

"I bow to your judgment, but wonder about the wisdom of the change. In Florence you serve your prince, who is indeed a wise and good monarch, but he is only a man, who may lend an ear to those envious of you, and withdraw his favor. Then, too, like every other prince, he has many interests. Today he may go about looking at the city of Florence through your spyglass, tomorrow more important things may require his attention and he will forget all about you and your lenses.

A prince is human. When he passes away, what is to prevent his heirs from casting you off? Think much on this."

Perhaps if Galileo had been able to talk to Sagredo about his desire to move to Florence, the change might never have been made. The rest of his life, then, might have taken a very different course.

In September, Galileo returned to his native Tuscany to live. Until such time as he could find a suitable house, he stayed with a good friend, Filippo Salviati. Many years later, Galileo was to honor these two friends, Sagredo and Salviati, by making them the chief characters in his most famous book.

Now that Galileo was free of all outside demands upon his time and energies, he set up his spyglass once more to explore the heavens. So far, he had not examined Venus closely. He focused his lenses on the brilliant planet.

To his great surprise, Venus appeared not as a shiny round object but as a crescent, like the new moon! He observed it for many nights and noticed a slow change taking place. The crescent became fuller, but the diameter of the entire planet seemed smaller. For about three months, he watched Venus become fuller and smaller, and finally show a round disk, with the smallest diameter of all:

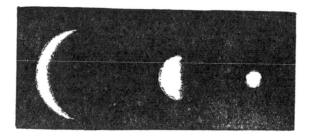

Then, slowly, it lost some of its fullness and gained in size.

Galileo became excited. The moon showed three *phases*—a series of changes to the eye—as it traveled around the earth, and these phases were due to its position as one looked at the part receiving light from the sun. Venus, too, showed phases. It, therefore, must also be receiving sunlight, and one saw these phases because of its position when one viewed the lighted part. But there was one great difference. The diameter of the moon did not appear to change with its shape, while Venus' apparent size did. What could account for this?

Since the moon was going around in a circle with the earth as its center, he argued, any time we looked at it, we saw it at the same distance from us. Suppose Venus was traveling around the sun, in a path between our earth and the sun, and receiving light from it. Then, when it was nearest the earth, we would see only a thin, long crescent. When its path took it farthest from us, it showed us a full face, but because of its greater distance, appeared smaller.

Venus, indeed, must be traveling around the sun, as Copernicus had stated over sixty years ago. The Polish astronomer had mentioned in his book that if our eyes were sharp enough, we would notice that Venus showed phases like the moon.

Copernicus was right! The sun was at the center of the solar system, and the earth and the other planets traveled around it. The proof Galileo had sought for so many years was found at last.

Why Venus shows us phases

XI

Galileo Revisits Rome

A LETTER FROM ROME REACHED GALILEO late in 1610. Father Clavius, chief mathematician at the Roman College, wrote him that the four moons of Jupiter had been observed in that city. This was the same Father Clavius who had befriended Galileo twenty-three years earlier when, as a penniless youth, he had visited Rome. The Jesuit priest mentioned in his letter that Galileo would receive a warm welcome should he decide to pay the city another visit.

A burning desire seized Galileo to quickly spread the news of his discovery of proof of the Copernican theory. If he could convince the Roman scholars, others would be impressed, too, for these churchmen were much respected by even the most worldly and daring.

The trip to the Holy City was planned for the end of December, but it had to be postponed until March. During the first winter after his return to Florence, Galileo suffered such a severe attack of arthritis that he had to remain in bed for most of that season. Physicians feared for his life; or that, even if he lived, he would surely be permanently crippled. But the astronomer's great will to live and to move about triumphed over his illness. Early in the spring, he was ready to leave for Rome.

Galileo took Rome by storm. Luxurious living quarters were reserved for him in the Embassy of the State of Tuscany. He was treated as an honored guest. Wealthy and influential noblemen welcomed him into their homes, and high church officials sought him out. Even the pope, Paul V, granted him an audience. This was indeed a triumphal return.

The head of the Roman College, Cardinal Robert Bellarmine, spent much of his time with Galileo discussing his discoveries. He himself had very little training in science but had taken the trouble to find out something about the new astronomy. He had asked the opinion of Jesuit professors of his college, and most of them spoke very favorably of Galileo.

However, Cardinal Bellarmine was troubled by some of Galileo's theories. For centuries the church had officially accepted and taught the science of Aristotle. It had taught, too, the theory of Ptolemy, which said that the earth stood still and the sun revolved around it. Should ignorant men have their faith in these teachings shaken, it was possible that they might begin to question the church in matters of religion, too. Cardinal Bellarmine asked Galileo to go slowly and carefully, and to speak of this new astronomy as a theory, not entirely proven, rather than as an absolute fact.

Galileo thought this over. The cardinal had a powerful influence over all of Catholic Europe, and his goodwill was very important. As far as he could, the astronomer would do as he suggested and spread the Copernican theory slowly and gradually.

One day in Rome, Galileo received an invitation to attend a meeting of the Lincean Academy—the Academy of the Lynx-Eyed. The founder of this academy, Count Federico Cesi, stated that he and the other members would be honored to have him join the Linceans.

Such an invitation was not to be treated lightly. Count Cesi was a learned young nobleman whose wealth made it unnecessary for him to work. Instead of idling his time away in frivolous pleasures, however, he sought the company of other scholars like himself. From among them, he very carefully selected a few of the best and asked that they join him in forming an academy. He proposed that they meet daily, whenever possible, and study science, mathematics, and astronomy.

Academies were not new in Italy. There were academies for the study of art, music, literature, and philosophy. They were not connected with any college or university. Rather, they were made up of men who wanted to choose their own subjects for serious discussion and study, without being hampered by any rules but their own. They encouraged artists, writers, and musicians, often helping them with money or by having their music played in public and their books printed. Being asked to join any academy was an honor, and being invited to membership in the scientific Lincean was one of the greatest.

Needless to say, Galileo accepted the invitation. Upon joining he was asked to add the word "Lincean" or the letter "L" after his name whenever he signed it, as a mark of honor to both himself and the academy. The members pledged themselves to use their influence and their money to help spread his scientific work. They would welcome an opportunity to print any of his future books.

During the first few days of membership, discussion centered around the spyglass. One of the members thought the name "spyglass" was awkward and did not really describe the instrument. They tried to think of a better name but could not. Finally, Cesi asked a friend of his, the Greek scholar Damascianus, if he could make up a name using Greek words. Damascianus coined the word "telescope," and so it is called to this day.

Galileo returned to Florence in June, pleased with the results of his visit to Rome. The name "telescope," which he liked, was officially accepted; he was a member of the Lincean Academy; he had renewed his friendship with Father Clavius, but best of all, many important church officials were impressed with his discoveries.

In his position as court mathematician, few official demands were made upon Galileo by the grand duke. He was really free to pursue his own studies. Sometimes he visited his daughters at the Arcetri convent, where they were very happy. Soon, he told them, little Vincenzo would come to Florence and live with their aunt Virginia. He spent days with his friend Salviati at his villa, where the two men found great pleasure in each other's company.

During the last week in September of that same year, two cardinals happened to be passing through Florence. The grand duke invited them to dinner and asked Galileo to attend. One of the cardinals, Maffeo Barberini, was enchanted by the astronomer's wit. He confessed that he himself wrote poetry and would like Galileo to read some of it. They became quite friendly during the dinner.

As the evening wore on, the talk turned to scientific discussion of floating objects. Aristotle was wrong, Galileo declared, and described his own experiments to prove it. Cardinal Barberini sided with Galileo, but the other cardinal, Ferdinand Gonzaga, opposed him. Cardinal Gonzaga refused to accept anything that contradicted Aristotle. Before the evening was over, Galileo decided to write a short book on the science of floating bodies and send it to the Lincean Academy in Rome to be printed. He thought, too, that he had made a good friend in Cardinal Barberini and an enemy in Cardinal Gonzaga. He felt now, more than ever, that the followers of Aristotle would someday do him great harm.

Though he was busy preparing his work on floating objects, which he planned to write in Italian for anyone to read, Galileo did not neglect his telescope. What he was especially eager to examine was the sun itself. In observing it, he noticed dark spots on the sun, irregular in shape and of different sizes. They gave no hint as to how they were caused or of what they were made. He watched them daily and soon discovered that they seemed to move across the sun. In letters to other astronomers about these sunspots, Galileo wrote that the movement of these spots presented two possibilities. Either they were material on or near the sun and moved around it, or—what was more likely—the sun rotated on its own axis, and the sunspots turned along with it.

Sunspots

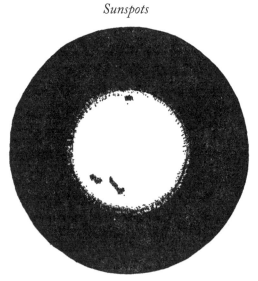

Galileo was not the first astronomer to observe and write about sunspots, but the letters he wrote about them were printed by the Lincean Academy and widely circulated. Those who opposed Galileo used these letters to try and show that the astronomer falsely claimed to be the discoverer of sunspots—and, in fact, that he was not quite the great astronomer he was said to be.

Thus the opportunity arose for those who were envious of Galileo's success to become bolder in their attacks on him. The followers of Aristotle, too, began to realize how much headway he was making in disproving beliefs held for so many hundreds of years.

Other signs of trouble began to appear. One of the most popular priests in Italy preached a fiery sermon against Galileo and the Galileists. His enemies were banding together and combining their efforts. Rumors reached Galileo that the Holy Office in Rome was looking into his teachings and that the cardinals were reconsidering their approval of him.

Galileo became alarmed. Disapproval of his teaching the Copernican theory could bring bad results. He wanted to know what was going on behind his back and how true these rumors were. Much as he dreaded traveling in the winter for fear of another attack of arthritis, he felt he must leave at once for Rome to answer his enemies.

Grand Duke Cosimo was sympathetic and helpful. He gave him letters to powerful people in Rome and again arranged to have him stay in the Tuscan Embassy. For as long as Cosimo lived, Galileo knew that in him he would have a protector.

He arrived in Rome in December 1615. The whole atmosphere of the Holy City had changed. A very different attitude toward the astronomer could be felt immediately.

Churchmen who had welcomed him five years earlier were now cold and unfriendly. Noblemen who had showered him with invitations were now hard to reach. Though the buildings and the streets were the same, it was hardly the same city for Galileo. His enemies had done their work well.

Count Cesi of the Lincean Academy, however, greeted him with open arms. He tried to explain the situation to Galileo.

Father Clavius had died, and his place was taken by a man less friendly to Galileo. Reports were being circulated that the astronomer was contradicting the Bible and that the church officials might even bring him before the dreaded Inquisition.

It is very difficult for us today to understand why this word struck terror in Galileo's heart. We are accustomed, whatever our religious beliefs, to have no fear of torture because of them. But in Galileo's time, things were very different in European countries, especially in most of Italy and in Spain. Democracy, as we know it today, did not exist then. At that time, whenever the government was under the strong influence of the church authorities in Rome, a person could be delivered to these church authorities for questioning about his beliefs. Often he would be tortured until he was willing to admit his errors and mend his ways. It did not matter that a person's body suffered briefly, for it was sincerely believed that, in the end, a faithful person would be saved forever. The court of inquiry was called the Inquisition, and innocent and guilty alike feared to be called before it.

Both the Tuscan ambassador and Cesi urged Galileo to be cautious. He had made many enemies among the followers of Aristotle who were looking for an excuse to denounce him before the Inquisition. His friends asked him, for the time being, to avoid all discussion of the Copernican theory. Spreading this theory seemed to be at the bottom of all his troubles.

In two months, Galileo managed to see many of his old friends among the church officials. He was able to convince them that he was a true son of the Catholic Church. Were not his two daughters novices at the Convent of San Matteo?

The Council of Cardinals at last handed down a decree to Galileo in February 1616. He must not teach or write about the Copernican theory as being absolute fact. As regards to Copernicus' book, *De Revolutionibus Orbium Celestium*, it was banned in its present form. When certain changes were made in it so that his theory was presented as an idea and not as a fact, it might be presented again.

Meanwhile, false rumors hostile to Galileo were being circulated. It was being said that he had appeared before the Inquisition and had been tortured; other rumors had it that he was imprisoned in a dungeon. To still these rumors, Galileo turned to Cardinal Bellarmine, who gave him a certificate clearing him completely. It stated that no fault had been found with the astronomer personally, nor had he been punished in any way.

Sadder, but much relieved of fear, Galileo returned to Florence. He prayed to be able to carry out the orders of the Council of Cardinals.

XII

Out of Retirement

AT THE AGE OF FIFTY-TWO, and at the height of his powers, Galileo had, in effect, been placed in retirement as an astronomer. Since he was forbidden to teach or write about the Copernican theory, there seemed little use in exploring the heavens any further. Whatever else he would discover would likely support the very theory he dared not spread.

Yet such an important part of his life had been given over to the study of the heavens that it was impossible to suddenly abandon it. Someday there might be a change, and the ban would be lifted. Should it come, he must be ready. Meanwhile, there was other work still to be done in mathematics and physics.

Shortly after his return to Florence, Galileo moved from the noise of the city to the Villa Segni, a small estate on the outskirts of the city. This was the grandest house he had ever owned. It had plenty of room for a small observatory, a beautiful garden, and a magnificent view of the surrounding countryside.

When he became tired of working indoors, Galileo

would go out and putter in the garden, work with his hands, prune trees, and experiment with growing different kinds of fruits. The pleasure he took in his garden helped a little to heal the hurt his spirit had suffered in Rome.

Sundays his chief joy was to visit his daughters at the Convent of San Matteo at Arcetri, where both had taken their vows as nuns. Virginia was now Sister Maria Celeste, and Livia, Sister Arcangela.

Maria Celeste was a kind, gentle young woman, always happy to see her father, ready to distract his solemn thoughts with stories of life at the convent. She worried a great deal about his health and constantly urged him to take care of himself. Sometimes Galileo would wonder which was the child and which the parent.

Arcangela, or Angela, as her father called her, was much more shy and withdrawn. It was difficult to know whether she enjoyed her father's visits or looked upon them merely as another duty to be performed. Try as he would to be impartial, Galileo could not help but favor Maria Celeste. He saved every letter she ever sent him. These letters later served as a source of much information about his private life in those years.

Vincenzo, his son, had come to live with Galileo's sister, Virginia. He fit in so well with his family of cousins that he was like a brother to them. The boy's education, however, was undertaken by his father. Daily Vincenzo came to his father's house, where he received instruction in mathematics and science. Before long Galileo realized, sadly, that his son had no great interest in these subjects. In fact, he had no great interest in learning of any kind. Mostly, he wanted to play with his cousins and let someone else worry about his future. The boy was a sore disappointment to his father. It looked more and more as though the only comfort of his old age would be Sister Maria Celeste.

Though the days seemed to pass quietly and peacefully, Galileo's brilliant mind was not at rest. He kept on turning over and over in his mind all kinds of ideas, hoping to find some outlet for his genius. Would it be possible, he wondered, to combine his knowledge of physics, mathematics, and astronomy with his skill at instrument making? He thought back on what he knew of the moons of Jupiter, and something struck him. There could not possibly be any objection to making practical use of their rotation around the planet.

He went to work with renewed vigor and was able to work out a method of locating ships at sea by observing Jupiter's moons. This method he offered to sell to the Spanish government. Spain might be particularly interested because of its many ships sailing over the ocean to its colonies in North and South America. The Spanish government promised to consider his method, but it must move slowly. The Italian astronomer must not become impatient if he did not hear favorably from them soon.

Illness could not have struck Galileo at a worse time than in the autumn and winter of 1618. While he was confined to his bed, three comets appeared in the sky. He could not get up to look at them either with the naked eye or through his telescope. Night after night the comets shone. Two were small, but the third was brilliant, with its long tail spread out like a magnificent bridal veil. Ignorant and superstitious people feared this to be a sign that the end of the world was at hand, for little was then known about these strange heavenly visitors.

Life had dealt Galileo a cruel blow in denying him the opportunity to observe this rare sight. Never again in his lifetime might another such as this appear. He would have to see it through the eyes of others.

The sick room became a meeting place of friends and one-time pupils. Various opinions were voiced about comets, and each was discussed at great length. Thus the dreadful months of painful illness were lightened by these kind and thoughtful men.

From these discussions, Galileo came to the mistaken conclusion that comets were burning masses of vapors freed from the atmosphere of the earth. He could not account for their tails and suggested that they might be due to imperfections in our sight. One of the visitors during his illness was a former pupil, Mario Guiducci, who in the spring gave two lectures on theories about comets. These lectures aroused so much interest that he decided to have them printed.

Guiducci was known to be a great admirer and follower of Galileo. Though the book on comets was printed by him, there was little doubt in the minds of most readers that the ideas expressed, and even many of the words, were those of the Master Astronomer himself.

Galileo's enemies could not resist this opportunity to strike at him again. A few weeks after the publication of Guiducci's book on the comets, another book appeared. It was called *Astronomical and Philosophical Scales on which Lothario Sarsi of Siguenza Measured the Theories of Galileo Galilei Expounded and Recently Published by Mario Guiducci at the Florentine Academy.*

There was no such person as Lothario Sarsi. The book was written by one of Galileo's strongest opponents, Father Grassi. In it, he claimed that Galileo was a fraud. Galileo had invented and discovered nothing new. Everything he had claimed to have done was either borrowed or stolen from someone else.

Such a violent attack upon himself could not go unanswered, but how to respond safely presented a difficult problem. Each word must be thought out carefully lest it be misunderstood. For three years since his return from Rome, Galileo had written nothing, for fear of antagonizing some of the church authorities, but silence was now unthinkable. Even if examining every phrase meant that it would take several years to write, in time, Father Grassi must be refuted.

Galileo's answer to the attack on him was interrupted by the sad death, in 1621, of his beloved friend and noble protector, Grand Duke Cosimo II. Now a ten-year-old heir, Prince Ferdinand, became head of the State of Tuscany. According to the law, Ferdinand's mother and grandmother took over the affairs of the duchy until the boy was old enough to rule by himself.

Not only did Galileo feel this to be a great personal loss, but his future was now no longer as secure as he had imagined. The boy, who was shortly to become his pupil, was completely dominated by his mother and grandmother. Though the ladies did not dislike Galileo, they were not interested in his science. For the sake of Cosimo's memory, the astronomer thought they would not dismiss him, but neither would they particularly favor him. What was more important, the two duchesses were friendly to some of the churchmen who did not like him.

"Many years ago my friend Sagredo wrote me from Venice that just such a situation might occur," thought Galileo. What a wise man Sagredo was!

Now, more than ever, he had to learn to watch his words. Slowly he proceeded with his answer to Father Grassi. Not until 1623 was he able to finish his little book. He called it *Il Saggiatore*, or *The Assayer*, and its title page explained its purpose. He wrote: "In which, with a most just and accurate balance, there are weighted the things contained in *The Astronomical and Philosophical Balance* of Lothario Sarsi of Siguenza."

Before the book was printed by the Lincean Academy, a wonderful piece of news reached Galileo. His friend and admirer, Cardinal Maffeo Barberini, had become pope. Upon election he had assumed the name of Urban VIII. What better way of congratulating the new pope than to dedicate his book to him?

In October 1623, *The Assayer* came off the press. For almost eight years Galileo had been silent. Eager to read what he had to say after all this time, people rushed to buy the book. Pope Urban VIII, pleased and flattered, let it be known that he would welcome a visit from his old friend.

Count Cesi wrote to Galileo that it might be a very good move if he came to Rome for a short stay. The new Pope had always been favorable to the astronomer, and he might possibly remove the ban placed on teaching the Copernican theory. Galileo agreed to leave Florence as soon as possible.

With the winter came another severe attack of arthritis. There was no denying it even to himself—he was getting old. At sixty, his attacks were more and more frequent and painful. Sometimes his sister Virginia took care of him; more often he hired a housekeeper. Sister Maria Celeste kept her father in good spirits with her cheerful letters, but Sister Arcangela seldom took the trouble to write. Neither was young Vincenzo any help to his father. The boy, now seventeen, still showed little interest in learning and had no sense of responsibility toward himself or anyone else. Even though he was ill, the father worried about his son's future. At the University of Pisa, the youth might possibly discover what he wanted to do for a living. Galileo enrolled his son as a student at the university where he himself, so many years ago, had come to study medicine.

By the time April arrived, Galileo was able to travel. Count Cesi wrote that he was being eagerly awaited in Rome. *The Assayer* was a great success, and it was time the astronomer came out of retirement.

Count Cesi was right. The silence and fear which had met Galileo in 1616 seemed to have lessened. Hope for lifting the ban grew stronger, and with this in mind, he asked for an audience with the pope.

The Holy Father greeted him warmly. They spoke of the past, of poetry, of literature, of mutual friends, of everything but what was dearest to Galileo's heart. Each time he introduced the topic of lifting the ban on teaching the Copernican theory, the pope managed to change the subject. Soon he realized that Pope Urban was a very different man from the Cardinal Barberini he had known. The responsibilities of his office had made him change his views on many things. The astronomer saw that personal friendship would not help him in matters affecting his writing and teaching.

Six audiences with the pope were granted. This was a special mark of favor from so busy and important a person. If only Galileo could get his permission to go ahead with a book he was thinking of writing!

Very carefully, he again introduced the matter of the Copernican theory to the pope. The Holy Father explained that there was no objection to writing it as a mental exercise, provided arguments both for and against the theory were given. However, the author must not draw the conclusion that the earth moved around the sun.

Galileo was well satisfied. If he used his words properly

in writing the book that was taking shape in his mind, he could present arguments for and against Copernicus. But he could do so in such a way that the reader himself could draw only one conclusion, namely, that the earth *did* move!

XIII
The "Dialogue" Brings Great Trouble

AN INTERESTING INSTRUMENT WAS SHOWN to Galileo while he was in Rome. It, too, was made of lenses and a tube, but this one did not enlarge things from a distance. When tiny objects were placed beneath the lower lens, they became greatly magnified. A thin hair looked like a heavy strand of wool, a tiny seed like a pea. Zacharias Janssen of Holland was credited with inventing this "compound microscope," and his fellow countryman, Cornelius Drebbel, had improved it.

Galileo brought one back with him to Florence. He took it apart and put it together again many times over, looking for a way to improve on it. Finally, by the end of August, he had done all he could and sent it back to friends in Rome with a letter telling of his improvements.

All the while that he was working to perfect the microscope, a part of his mind was elsewhere. Little by little, a great book was taking shape in his brain. He would call it *Dialogue on the Great World Systems*, and it would be written in Italian for any educated person to read. Moreover, it would not be like the dull, complicated science books that most people avoided, but it would be made up of witty conversations carried on by three men.

The *Dialogue* would take place in Venice. The three gentlemen would meet for several days and discuss the two main theories of astronomy. One man, called Simplicio, would be a believer in the Ptolemaic theory that the sun goes around the earth. The second, Salviati, would be a follower of Copernicus, and the third, Sagredo, would ask questions of the others to keep the discussion going.

The work on the *Dialogue*, of course, would be slow, but interesting. Even if it took many years to finish, it did not matter as long as, in the end, the book would be just right.

Meanwhile, Grand Duke Ferdinand II of Tuscany was old enough to begin studying mathematics. Galileo attended his duties at court and found the young duke a very fine pupil, just as his father Cosimo had been. Ferdinand, too, showed great liking and respect for his teacher and once more offered him the friendship and protection of the Court of Tuscany.

If only his own son Vincenzo were different! But Vincenzo was an idle spendthrift, always asking his father for money so that he could live like a wealthy prince at the University of Pisa. One day Galileo wrote in an angry letter to him: "Learn to live within your allowance. At your age I did not have as many groats to eat as you have gold florins to squander."

Vincenzo did not even take the trouble to answer. Only when he needed money again did he write to his father. At last, after receiving a degree at the university, a position as a clerk was offered him. He accepted it with poor grace, for in his opinion, it was not grand enough for a Galilei. But at any rate, he was earning some money of his own, and he wanted to get married.

Galileo gave him his blessing, hoping that marriage would cause him to settle down. When Vincenzo and his wife presented him with a grandson, whom they called Galileo in his honor, the aging scientist was pleased to know that his name would not die with him.

It took Galileo almost five years to finish the *Dialogue*. Age was not slowing his mind, but it was having its effect on his body. Many times he found it necessary to lay aside his pen and take to his bed because of illness. At last, in 1630, he was ready to take his completed book to Rome to receive permission to print it.

In those days, no good member of the church bought a book unless permission to print it—the *imprimatur*—had been granted by church authorities. Nor would a printer take a chance and invest money in any book unless it had received such an *imprimatur*. Hence Count Cesi of the Lincean Academy urged Galileo to receive official permission as soon as possible so that the printing could begin.

The secretary in charge of the *imprimatur* granted permission to print without examining the *Dialogue* as closely perhaps as he should have. All was ready to start putting the book into production when tragedy struck. Count Cesi died of the plague, which was then striking a large part of Italy.

Galileo mourned his friend, the founder and president of the Linceans. With Cesi's death, the Academy disbanded. Now that there was no one in Rome to supervise the printing of the book, the author took it back with him to Florence to have the work done there. But first, permission to print in that city had to be obtained.

The plague was spreading through Europe like wildfire. No one knew its cause, prevention, or cure. One day a man was seen walking through the streets in fine health, the next day he was dead. People avoided each other and, as far as possible, shut themselves into their homes so as not to have contact with anyone carrying the "Black Death," as it was called.

Communications between Rome and Florence were suspended until the plague subsided. Because of this, permission to print the *Dialogue* in Florence had to wait until contact between the two cities was resumed once more.

The Villa Segni, Galileo's home on the outskirts of the city, had become too large for the aging man to take care of. Besides, the walk to his daughter's convent was too long and hilly for his aching legs. He asked Sister Maria Celeste to inquire about a small house for him in Arcetri so that he could be nearer to them.

Just the right house became vacant in 1631. It was smaller than the Villa Segni but sufficient for his needs. Surrounding it were vineyards which Galileo loved to cultivate, and better yet, he was almost within sight of the Convent of San Matteo. Quietly tending the grapevines, doing small chores at the convent, fixing its clock, he passed the days while waiting for the plague to ease and mail to come.

Late in 1631, the good news arrived. Galileo rushed into Florence with his handwritten *Dialogue* and the *imprimatur* to the printing house of Landini. In February 1632 the book was ready for sale.

Suddenly, in August, an order came from Rome to stop all sales of the *Dialogue* immediately. Whoever was responsible for granting permission to print it had not read it as carefully as he should have. Now that it was being circulated and other churchmen had read it, they were aghast at what had been done. Galileo had followed the pope's

DIALOGO
DI
GALILEO GALILEI LINCEO
MATEMATICO SOPRAORDINARIO

DELLO STVDIO DI PISA.

E Filosofo, e Matematico primario del

SERENISSIMO

GR.DVCA DI TOSCANA.

Doue ne i congressi di quattro giornate si discorre sopra i due

MASSIMI SISTEMI DEL MONDO
TOLEMAICO, E COPERNICANO;

Proponendo indeterminatamente le ragioni Filosofiche, e Naturali tanto per l'vna, quanto per l'altra parte.

CON PRIVILEGI.

IN FIORENZA, Per Gio:Batista Landini MDCXXXII.

CON LICENZA DE' SVPERIORI.

Title page of the original Dialogue on the Great World Systems, *printed in 1632 (Courtesy Rare Book Collection, New York Public Library)*

Illustration facing the title page of the Dialogue.
Here Simplicio is asking questions of Sagredo, dressed as Ptolemy, and of Salviati, dressed as Copernicus.
(Courtesy Rare Book Collection, New York Public Library)

suggestion and presented both theories concerning the movement of the sun and earth, but only a fool could be blind to his real intent. Salviati, the Copernican, was by far the cleverest of the three characters. His arguments were so good that anyone reading the *Dialogue* with even half an open mind could hardly fail to be convinced that the earth moved around the sun.

Galileo's enemies laid their plans carefully. Once and for all this doubter, this dangerous "seeker after the truth," as he called himself, must be silenced. Even his friends within the church became alarmed at the possible effect acceptance of the Copernican theory might have at a time of such trouble in the world. Plagues, revolutions, and revolts against the authority of the church in other parts of Europe made the people restless and uncertain.

For the sake of peace, it would be much wiser not to permit any new ideas to spread just now. Moreover, it looked as though the astronomer had disobeyed the order of 1616 not to write about or teach the Copernican theory.

Once more Galileo was gripped with fear. His terror was well founded, for, in October, two months after all sale of the *Dialogue* was stopped, he was summoned to appear before the dreaded Inquisition in Rome.

At this piece of news, the sixty-nine-year-old man collapsed. Both shock and arthritis combined to make him so ill that the doctors did not think he would recover. Grand Duke Ferdinand dispatched a messenger to Rome asking for a delay in the trial, and when medical certificates were produced, such a delay was granted.

In February of 1633, Galileo finally came to Rome, a tired, sick old man. He still had some good friends, however. Grand Duke Ferdinand gave him rooms in the Medici Palace, where the Tuscan ambassador lived, and also used his influence to ask that the seventy-year-old scientist be treated gently. Galileo was grateful, for he'd had visions of being imprisoned in the dungeon of the Inquisition until his trial.

The embassy became his home for more than four months while material was being gathered for his trial. In that time, he was questioned unofficially about his beliefs, his actions, and his work. Finally, he was ordered to appear before the Inquisition at the Castel Sant'Angelo.

What would happen to him? Would he be cast into a dungeon and tortured until he admitted his faults? Galileo could not still his dread. To his great surprise, upon his arrival at the headquarters, he was conducted to a comfortable suite of three rooms in the Castel Sant'Angelo, and a servant from the embassy was permitted to stay to help take care of him. The Tuscan ambassador, Niccolini, sent special food to the sickly old man for the four days he remained there.

During these four days, he was again questioned closely about his beliefs and his works. At night he was conducted back to his rooms, where he spent sleepless nights tossing on his bed, wondering what course he should take.

The questioning showed him that he would be asked to deny the truth of the Copernican theory. To insist that he still held strongly to his beliefs might mean torture and painful death, yet to deny them would be looked upon by his followers as cowardice.

Whether he lived or died, he knew the truth could not be crushed. If the Copernican theory were indeed true, as he believed, nothing anyone said could bury it. No law passed by men could make the earth stand still if it really moved.

In despair Galileo decided he could not fight any longer! Let younger and stronger men carry on where he left off.

There were enough of his pupils scattered throughout Europe to spread the truths he had discovered. He would obey the orders of the Inquisition.

Before the Council of Cardinals, he brought himself to confess: "I do not hold, and have never held, the opinions of Copernicus since the order was given me to abandon them. In any case, I am in your hands. Do with me as you please."

Galileo before the Inquisition

On June 22, 1633, weak with age, illness, and shame, Galileo was brought to the Church of the Santa Maria Sopra Minerva to hear this sentence passed upon him:

First, the *Dialogue on the Great World Systems* was prohibited.

Second, for three years he must recite once a week the seven psalms of repentance.

Third, he was to be imprisoned in his own home for an indefinite period of time, for as long as it pleased the council.

A broken man, hating himself for his own cowardly behavior, he was released in custody of his old friend, Cardinal Piccolomini of Siena. He was permitted to stay at the cardinal's home until well enough to return to Arcetri.

In his own heart, Galileo knew that he had confessed falsely. *The earth did move!*

XIV

Truth Cannot Die

THE GRAPES IN HIS VINEYARD had long since ripened and dried out on the vines when Galileo returned to Arcetri from Siena. December winds whistled through the chimney of his house, but he did not mind. After eleven hard months, it was good to be back home.

It occurred to Galileo that there was some consolation in not being permitted to leave his house except to visit his daughters at the convent. In his shame, he would not have to face people and try to explain his actions. Neither would he be embarrassed by seeing friends turn away from him as he walked along the streets of Florence.

Sister Maria Celeste's joy at seeing her father back, safe and sound, knew no bounds. She comforted him with kind words of understanding. Galileo was distressed at her appearance, for though she had never been very strong, she was thinner and paler than he had ever seen her. Unceasing work in the convent apothecary during the years of the plague were showing their effects on her. Now the father urged the daughter to take care of herself, as she used to urge him in years past.

Confined to his house and vineyard, the great scientist brooded about the past. If he had stayed in Padua, the Republic of Venice would never have permitted the Inquisition to touch him. His book might have been forbidden, but he himself would not have been brought to trial.

But it was useless to regret what was past. His problem at present was to find work to fill his days and to re-establish his good name. Aimlessly he walked about his house looking for something to do. He thought of all the notes he had made on his scientific work since early youth; they were still stored away in chests. Arranging them in some order would make a good start. He set about with a will to drown his sad thoughts in hard work.

Instead of arranging the papers according to date, he put them into piles according to subject, out of curiosity to see what he had accomplished in each. When at last he was finished, he examined the material before him, and a familiar thrill of excitement shot through him. There was enough material for a book on the science of physics, and work for the rest of his life.

Spring came, and with it great sorrow. The Mother Superior sent a message to Galileo saying that Sister Maria Celeste had died. Her father mourned her, not just as a loving daughter but as a comforting friend. Now he had no one, really. Sister Arcangela seldom came out of her nun's cell to see her father, and Vincenzo came to visit only when he wanted money.

The lonely old man took the death of his favorite daughter so much to heart that he became very ill. His son did not take the trouble, even then, to visit his father often. A trusted housekeeper nursed Galileo back to health. His recovery was speeded by hopeful reports from the outside world.

"The truth cannot be crushed," Galileo had said to himself two years earlier. "No laws passed by men can bury it." His prediction was coming true. Word came to him that the *Dialogue on the Great World Systems* had been translated into Latin and printed in Strasbourg. Many people were buying it out of curiosity to see what was in this forbidden book, but even more were buying it to study the Copernican theory. Ships carried copies of the book to all parts of the civilized world. A translation into English was being prepared, too. In trying to ban the theory of the earth's movements, the churchmen had only succeeded in calling attention to it. More and more, it was gaining acceptance.

All these reports gave Galileo courage to carry on with his work. Little by little, friends who had drifted away came to visit or wrote him letters. Though he could not leave his house and grounds, life was no longer as empty as when he had returned from Rome in disgrace.

Vincenzo's family increased. He now had three children and needed money more than ever. Hard work was not to his liking as long as his father was there to hand out money. But because there was not so much money to be given him these days, he thought of a good way of getting some.

He reminded the old scientist of his discovery of a method of locating ships at sea. After many years of waiting, Spain some time ago had decided not to buy the plans. Galileo had forgotten all about them, but at Vincenzo's insistence, he offered the plans to Holland.

The Netherlands' government was anxious to have them but was unable to pay him for their use, though they were more than willing to do so. No one from Holland, a Protestant country, was permitted to visit Galileo. Rather than risk further trouble, he let the whole matter drop.

Vincenzo finally had to go to work. He was trained as a lawyer and found a good position where he could put his knowledge to use. For a while, he managed to support his family by himself.

The characters of Salviati, Sagredo, and Simplicio of the *Dialogue* came back to life in Galileo's new book. Simplicio disappeared near the beginning, but the others carried on in a discussion of the new science, mostly in Latin. Galileo called this book *Discourses on the Two New Sciences*. It was a complete, carefully written book on physics which summarized the work of his lifetime.

Among scientists, this is considered his greatest work. While the *Dialogue on the Great World Systems* made the Copernican theory popular in its own way, this new book did far more. It opened the eyes and minds of men to a new kind of science by the experimental method. The *Discourses on the Two New Sciences* was the first great book on modern physics.

Knowing that anything he wrote would never again receive the *imprimatur*, and that no one in Italy would print a book without it, Galileo sent the manuscript out of the country secretly in 1637. It was printed in Leyden, Holland, in 1638, but the author was never to see his book in print.

By the middle of 1637, Galileo noticed he was having trouble reading his own notes. Eyeglasses did not improve his vision, and, frightened at what this could mean, he sent for a doctor. To his despair, he learned that one eye was completely blind and that the other was becoming infected, too. Prompt treatment might save it, but doctors were not anxious to make frequent trips in winter to Arcetri.

The old man wrote a letter to Rome, asking permission to leave his house for medical care in Florence. He was

DISCORSI
E
DIMOSTRAZIONI
MATEMATICHE,
intorno à due nuoue scienze

Attenenti alla

MECANICA & i MOVIMENTI LOCALI,

del Signor

GALILEO GALILEI LINCEO,

Filosofo e Matematico primario del Serenissimo
Grand Duca di Toscana.

Con vna Appendice del centro di grauità d'alcuni Solidi.

IN LEIDA,
Appresso gli Elsevirii. M. D. C. XXXVIII.

Title page of Discourses on the Two New Sciences, *printed in 1638.
(Courtesy Rare Book Collection, New York Public Library)*

allowed to do so and moved into Vincenzo's house for treatment. The doctors, however, could do nothing for him. When a copy of the *Discourses on the Two New Sciences* was placed in his hands, he was completely blind.

Now he lived in total darkness, but his mind was still alert. Galileo wanted to work on additions to the book, and a house with three lively children always underfoot was hardly the place for a blind man. He asked to be taken back to his own home in Arcetri. There, among the familiar things he could feel and smell, he would be happier than at Vincenzo's.

A servant and a housekeeper went along with him. Gradually he learned to tap his way through the rooms and among the vines in his vineyard. Vincenzo came several times a week to act as his secretary and to keep

Vincenzo Viviani

him amused with the gossip heard in Florence. His father smiled to himself at the young man's sudden thoughtfulness, for he knew his son well. He had not much longer to live, and Vincenzo wanted to be mentioned handsomely in his father's will.

One day a visitor came to Galileo with a request. "Master, may I stay here with you and take care of you? I ask nothing but to be allowed to hear you talk to me of your new science."

The man was Vincenzo Viviani, who had once been the old man's pupil. He loved the old scientist. When he heard of his teacher's plight, he came and offered his services. Viviani moved into the house at Arcetri and served as secretary, friend, and good companion.

Evangelista Torricelli joined the household in 1641. He, too, had been a pupil of the great scientist. Torricelli had

Evangelista Torricelli

Last days at Arcetri

come to learn more from his old professor, as well as to tell him how successful his new book was.

They talked of many scientific inventions. Torricelli mentioned the air thermometer and its weaknesses, and Galileo suggested what the trouble might be. As a result of these talks, Torricelli got his idea about the mercury barometer.

The three men lived a quiet life in the little house in Arcetri. It hardly seemed possible that from this tiny corner of Italy, a whole new science had been sent forth into the world.

Winter brought feverish illness to the great master. No power on earth could help him in his weakened old age. Wearily he breathed his last on January 8, 1642.

Galileo's long struggle to "find the truth and proclaim it" was over, but he left behind him the means of going on where his own work ended. He will never die. Every time we look through a telescope or do an experiment in science, he smiles at us with approval.

XV

Important Dates in Galileo's Life

1564—February 15: Galileo born in Pisa, Italy

1581—September: Entered the University of Pisa as a medical student

1583—Discovered the laws of the pendulum

1585—Left the University of Pisa to return home to Florence

1587—Paid first visit to Rome, where he made friends with many important people

1589—Appointed professor of mathematics at the University of Pisa

1591—Experimented on falling bodies from the Tower of Pisa

1592—Appointed professor at the University of Padua

1594—Became ill with first attack of arthritis

1600—Daughter Virginia born

1601—Daughter Livia born

1605—Spent summer teaching Cosimo, Crown Prince of Tuscany

1606—Son Vincenzo born
Published book explaining the use of the proportional compass

1609—Summer: Constructed his spyglass and was the first one to use such an instrument to study the heavens
Autumn: Studied the moon through his spyglass and decided it was shining by light it received from the sun

1610—January: Discovered the four moons of Jupiter
March: Published *The Messenger of the Stars* telling of his discoveries by means of the spyglass
Autumn: Appointed as court mathematician to Cosimo II: Grand Duke of Tuscany, and returned to live in Florence
Discovered the phases of Venus and saw this as proof of the Copernican theory

1611—Visited Rome and hailed as a great astronomer
Joined the Lincean Academy
Spyglass given the name "telescope"

1614—Daughters entered the convent at Arcetri as novices

1615–1616—December to February: Revisited Rome; told to stop teaching the Copernican theory that the earth moved around the sun

1624—Visited Rome again; spoke to the new pope, a friend of long standing, and was told he might write about the Copernican theory if he would present arguments for and against it

1631—Moved to house at Arcetri

1632—February: *Dialogue on the New World Systems* published
August: *Dialogue* banned

1633—February to June: Examined by the Inquisition in Rome
June: Denied he believed in the Copernican theory; sentenced to house arrest at Arcetri for the rest of his life

1637—Became blind

1638—*Discourses on the Two New Sciences* published in Leyden, Holland; this was the first book on modern physics

1642—January 8: Died at Arcetri

More books from The Good and the Beautiful Library

Ladycake Farm
by Mabel Leigh Hunt

The Lost Kingdom
by Chester Bryant

Tiger on the Mountain
by Shirley L. Arora

Girl with a Musket
by Florence Parker Simister

goodandbeautiful.com